METEOROLOGICAL MONOGRAPHS

VOLUME 25   NOVEMBER 1994   NUMBER 47

MESOSCALE MODELING OF THE ATMOSPHERE

Edited by

Roger A. Pielke
Robert P. Pearce

American Meteorological Society
45 Beacon Street, Boston, Massachusetts 02108

ISBN 1-878220-15-2
ISSN 0065-9401

Published by the American Meteorological Society
45 Beacon St., Boston, MA 02108

Printed in the United States of America
by Lancaster Press, Lancaster, Pennsylvania

This monograph is dedicated to the memory of

Professor Jehuda Neumann

(1915–1993)

in recognition both of his outstanding pioneering contributions to mesoscale modeling of the atmosphere and his most valuable services to meteorological research in the U.S. Army as a member of its Advisory Panel in Mesoscale Meteorology for over 20 years.

TABLE OF CONTENTS

PREFACE

The past few years have seen an upsurge in mesoscale meteorology and, in particular, in the development of mesoscale models. Such models, many incorporating highly detailed physics, are now used extensively in experiments to improve understanding of mesoscale phenomena such as fronts, severe local storms, land and sea breezes (one of their earliest applications), tropical cyclones, flow around hills (including so-called large-eddy simulations), and associated effects such as the spread of pollutants. These models are also being used increasingly in operational forecasting, some as "stand-alone" models, others nested within synoptic- (or even global-) scale models. An example of the latter is the "unified model" currently in operational use at the U.K. Meteorological Office; another is the tropical cyclone prediction model at the National Hurricane Research Center, Miami, Florida.

The American Meteorological Society book *Mesoscale Meteorology and Forecasting,* edited by Peter Ray and published in 1986, is perhaps the most up-to-date comprehensive account of mesoscale meteorology currently available. It not only covers the basic theory of the whole range of mesoscale phenomena but also includes substantial sections on observing systems, mesoscale modeling, and short-range forecasting techniques. However, during the eight years that have elapsed since its publication, modeling developments have led to a focusing of attention on those particular aspects of mesoscale systems that models, particularly those in operational use, fail adequately to predict.

It therefore seems appropriate to publish a new monograph devoted to an up-to-date review of these topics. Also, since a comparison was recently carried out of the performances of four models using an observational dataset—the United States Army's Project WIND (Wind in Non-uniform Domains)—performed specifically for the purpose of model validation, a brief account of the results of these comparisons will be presented to illustrate the nature of the problems that mesoscale modelers still need to address.

Chapters 1–6 of this monograph provide an overview of the status of several components of mesoscale modeling that are of particular importance to the accurate simulation of the Project WIND observations.

The United States Army's Atmospheric Research Laboratory (ARL) at White Sands Missile Range, New Mexico, has been involved in mesoscale model applications for many years. The army requires such models to provide weather forecasts in any part of the world in which it is engaged in military operations. The At-

mospheric Science Laboratory (ASL), ARL's earlier designation, recognized in the mid-1980s that the models it was developing—ranging from the mesobeta (100 km) to the microscale (10 m)—needed an observational database of appropriate resolution if they were to be properly tested. It therefore designed for this purpose and carried out during 1985–87, under the direction of Mr. Ron Cionco, a field observation program in northern California called Project WIND. The observations were concentrated in four periods, each of about two weeks and centered on a different season of the annual cycle. The vast quantity of data obtained has now been almost completely quality assured and processed into an accessible form on computer tape or disk. Details of the field program and dataset are given in chapter 8.

The army's Mesomet Panel suggested in 1990 that advantage should be taken of the availability of processed subsets of the Project WIND dataset to invite modeling groups in the United States and Europe to take part in a model comparison project using data from the summer (phase I) and winter (phase II) periods to initialize the models and carry out 24-h forecasts on a 41×41 horizontal (5 km) grid. Details of the project implementation and of the requests made by the modelers are given in chapter 7.

Four groups participated and forwarded their results to ASL for analysis and presentation at a mesoscale modeling workshop held in El Paso, Texas, in June 1992. A selection of the results from each of the models is presented in chapters 9–12. Finally, a statistical analysis of some of the results is provided in chapter 13, and a commentary on the performance of the four models is offered in chapter 14.

Publication of this monograph is supported under Contract DAJA 45-90-C-0009 awarded to the University of Reading, United Kingdom by the U.S. Army European Research Office. The editors also wish to acknowledge with gratitude the support provided by the staff of the Atmospheric Research Laboratory, White Sands, New Mexico, without which the extensive data analysis involved in the mesoscale model comparison project described in chapters 7–14 would not have been carried out. Thanks are due, in particular, to Mr. Ron Meyers of ARL for also undertaking the task of collating the descriptions of the four models involved in the project and which constitute the appendices of chapters 9–12.

Roger A. Pielke
Robert P. Pearce
Editors

PART I

Aspects of Mesoscale Modeling

Chapter 1

Initial Conditions and Boundary Conditions

KEITH D. SASHEGYI AND RANGARAO V. MADALA

Naval Research Laboratory, Washington, D.C.

1.1. Introduction

High-resolution limited-area models have been successfully used for both the simulation and prediction of regional and local mesoscale weather events (Anthes et al. 1989). This success can depend on accurately specifying the initial state of the atmosphere and its variation in time along the lateral boundaries of the model domain. Synoptic-scale global or regional analyses produced at operational weather centers do not provide sufficient resolution for the initial conditions required by a mesoscale model. The observations themselves must be used to enhance a regional analysis or to correct an initial-guess field produced by the mesoscale model itself (e.g., by integrating the model from an earlier operational regional analysis). Often the data themselves are of insufficient resolution to effectively enhance a global- or regional-scale analysis alone, and the data must be used in combination with the high-resolution mesoscale model (Anthes et al. 1989). To provide the lateral boundary conditions for the mesoscale model, forecasts from an operational global or regional model are usually used. For such an operational forecasting system, the forecasts are often too infrequent to provide sufficiently accurate lateral boundary values for the high-resolution mesoscale model. In this case it is useful to nest the high-resolution mesoscale model inside a limited-area model. This provides a link between the mesoscale model and the sparse data and infrequent boundary values provided by forecasts from a global or regional model.

The components required to produce the initial conditions for a mesoscale model can be classified as (i) quality control, (ii) objective analysis, (iii) initialization, and (iv) assimilation. To utilize the data, some form of quality control is initially required. Any unreliable data that may be inconsistent with some initial-guess field, are not supported by neighboring observations, or are of a scale too small to be resolved by the analysis scheme (and can produce aliasing into the larger scales) need to be removed. To provide the initial conditions for a mesoscale model, the observed data must then be interpolated to the model grid points or some lower-resolution analysis grid if the observed data are too sparse. Objective analysis describes the methods used for such interpolation. Errors in the analysis (due to observational errors and unresolvable scales of motion) and inaccuracies in the model physics give rise to imbalances between the wind and mass fields. Such imbalances are removed during numerical integrations of the model by inertia–gravity wave oscillations. The purpose of initialization prior to integrating the model is to prevent such gravity wave noise contaminating the forecast particularly in the first 12 h of the forecast of the mesoscale model. When insufficient data are available on the scale required by the numerical model, the model can be used to provide the time continuity between the observation times and generate the spatial variability on the scales resolved by the model. Assimilation describes the process by which the data are introduced into the model to create a continuous description of the atmosphere, which has been updated by the data. During a period of assimilation, the model can spin up mesoscale features forced by variable terrain and surface heat distributions. Our interest in this paper is in generating more accurate initial conditions for starting a subsequent forecast with the model by assimilating the data during a period of time prior to the start of the forecast run.

An excellent description of the above aspects of the initial conditions can be found in Daley's (1991) recent book and its review by Harms et al. (1992a). In this paper, we will discuss the above issues as they relate in particular to high-resolution limited-area modeling. We will discuss in the following sections each of the four components listed above for the initial conditions. The influence of the methods for prescribing the lateral boundary conditions for high-resolution limited-area models will also be discussed in section 1.6 but in less detail. Conclusions with some discussion of possible future areas of research will follow at the end of the chapter.

1.2. Quality control

The data available for use in regional and mesoscale models are of varied quality and some form of verification of the data is required before use. The upper-air soundings from the rawinsonde network archived at national centers are not without problems (Schwartz and Doswell 1991). At operational weather centers, automated methods have been developed for quality control of the data (see, e.g., DiMego 1988). Initially, the soundings such as those from the rawinsondes are checked for hydrostatic consistency and possibly smoothed or interpolated in the vertical to be more representative of the model vertical resolution. In some current methods of quality control at operational centers, errors in the geopotential and temperature in the soundings from many sources are able to be corrected at this stage (Collins and Gandin 1990). The observations are then compared to an initial background guess field that is a forecast field generated by an earlier run of an operational model. In a "gross" check, observations with large deviations from the first guess are removed, and in a "buddy check," observations not agreeing with neighboring observations are removed. Close observations are replaced by an average "super" observation (hereafter superobs) in order to reduce the processing load on the analysis scheme and to keep some analysis schemes stable. Isolated observations that cannot be verified by the above procedure are also usually eliminated. An important result of the above procedures is that observations of mesoscale features that are at scales not resolved by the analysis grid are removed to prevent their aliasing to the larger scales. In regions of little data, bogus data may be added to include weather features missed by the observations or to merge with a larger-scale analysis outside of the observations of interest. Further details on quality control methods may be found in the excellent review by Gandin (1988) and the sources mentioned therein. More theoretical aspects of quality control are described by Lorenc and Hammon (1988).

1.3. Objective analysis

The most commonly used methods to interpolate the meteorological observations to a model grid or a separate analysis grid are based on using a linear combination of the observational data to obtain values at each grid point. Usually the deviations of the observational data from an initial guess or background field are interpolated instead of the data themselves. If ϕ_a is the interpolated (analyzed) value at the grid point and ϕ_b is the value of the background or first guess at the grid point, we can write

$$\phi_a = \phi_b + \sum_{i=1}^{m} w_i(\phi_{o,i} - \phi_{b,i}), \quad (1.1)$$

where $\phi_{o,i}$ are the observations, w_i are the weights for each observation, $\phi_{b,i}$ is the value of the background at the observation point (which is usually derived by a polynomial interpolation from ϕ_b), and m is the number of observations. The background field can be a climatological mean field or a forecast field produced from an earlier analysis. The method can also be repeated in an iterative fashion, so that the background field is updated by the latest analysis after each iteration or pass; that is,

$$\phi_a(n + 1) = \phi_a(n) + \sum_{i=1}^{m} w_i[\phi_{o,i} - \phi_{a,i}(n)], \quad (1.2)$$

where $\phi_a(n)$ is the value of the analysis at the grid point and $\phi_{a,i}(n)$ is its value at the observation location after the nth iteration.

The large-scale atmospheric flow is largely in (quasi) geostrophic balance, and the analysis scheme can be generalized to include such a dependency between the meteorological variables. In such a multivariate scheme, the ϕ_a, $\phi_{o,i}$, and ϕ_b are then vectors representing several variables (such as the horizontal wind components and the geopotential), and the weight coefficients become matrices of coefficients linking the variables. Further dynamical constraints can also be satisfied by applying variational methods, as introduced by Sasaki (1958, 1970a). A concise review of these objective analysis methods and others can be found in Schlatter (1988). Further details can be found in Thiebaux and Pedder (1987) and Daley (1991). The common linear methods above can be divided into those methods where the weights are empirically prescribed and those where they are derived by statistical considerations.

a. Empirical interpolation

In the empirical methods, the weights are prescribed empirical functions that decrease with the distance of the observation from the grid point. Only those data that fall within a certain critical distance of the grid point (called the radius of influence) are used. The weights are usually normalized by the sum of the individual weights used for each of the observations in the interpolation at the grid point. Usually, the iterative method, which is commonly called the method of successive corrections, is used. The successive correction method was first introduced by Bergthorsson and Doos (1955), and different versions have been developed by Cressman (1959), Barnes (1973), and others. Various derivatives of these earlier methods have been widely used for the analysis of mesoscale systems in the research community (e.g., Koch et al. 1983; Maddox 1980; Benjamin and Seaman 1985; Achtemeier 1989). By using repeated passes in these methods, successively smaller scales of the motion are analyzed with each increasing iteration [see, e.g., chapter 3 in Daley (1991)]. In these successive correction methods, the interpolated values at the data locations will eventually converge to the data values themselves. Since the data have some error, the number of iterations used is limited by some criterion, such as analyzing for only those scales that are resolved by the data network (Koch et al. 1983). After one or more iterations, the width scale of the weighting function is usually also reduced to speed the convergence of the scheme for the smaller scales being analyzed.

In the schemes of Bergthorsson and Doos (1955) and Cressman (1959), the observed winds are used to provide estimates of the geopotential gradient at the wind locations using the geostrophic relation. The observations of geopotential $\phi_{o,i}$ at the same locations in (1.1) or (1.2) can then be replaced by values extrapolated to the gridpoint locations,

$$\phi_{o,i}^* = \phi_{o,i} + (\mathbf{r} - \mathbf{r}_i) \cdot \nabla \phi_{o,i}, \qquad (1.3)$$

where $\nabla \phi_{o,i}$ is the gradient derived from the horizontal wind $\mathbf{V}_i$ at the observation point using the geostrophic relation. Here, $\mathbf{r}$ and $\mathbf{r}_i$ are the position vectors of the analysis gridpoint and observation location, respectively. To provide the feedback to the wind field, the resulting change in the geopotential gradient can be used to compute a geostrophic wind correction to be added to the wind field (following, e.g., Hayden 1973; Kistler and McPherson 1975).

In using many mesoscale models over small geographic regions, the upper-air observations are few and far apart, and these models have been started usually from a single atmospheric sounding. Segal and Pielke (1981) used two radiosonde observations to analyze the initial spatial variation of the wind over the Chesapeake Bay region, and a superior forecast was obtained when compared to using a single sounding. In their method no background was used in (1.1), and the weighting functions were defined by the inverse of the distance squared. In other cases a higher density of observations is available over a smaller region of the model or analysis domain, compared to the observations over the remainder of the domain. In using the above scheme to analyze the data over the whole domain, the observations in the regions of high data density are given excessive weight compared to observations in the areas of low data density. For example, in this case the number of observations must be reduced in the regions of higher density by averaging or eliminating some of the observations. Further passes with the analysis scheme on a finer-resolution grid can then be easily added in the regions of higher observational density. Such a scheme was described by Shi et al. (1991), who applied the Barnes scheme for high-density dropwindsonde data in the region of a hurricane. For the initial synoptic-scale analysis, the data in the hurricane core were removed by the quality control, correctly preventing the aliasing of the smaller-scale hurricane into the large synoptic scale. By then adding a finer-resolution grid in the region of the hurricane, the hurricane was successfully merged with the synoptic analysis.

b. Statistical interpolation

The most widely used method in the operational community for the analysis of synoptic-scale flow is the method of statistical interpolation, better known as optimal interpolation. With optimal interpolation—first applied by Eliassen (1954), Gandin (1965), and Eddy (1964) to meteorological data—the mean-square analysis errors are minimized. The analysis, background, and observation errors are usually defined as deviations from true values in the method. For the variable ϕ used in (1.1), defining the errors by $e_a = \phi_a - \phi_t$, $e_b = \phi_b - \phi_t$, and $e_o = \phi_o - \phi_t$, where ϕ_t are the true values, then an expression for the mean-square analysis error $E_a^2 = \langle e_a^2 \rangle$ is

$$E_a^2 = E_b^2 - 2 \sum_{i=1}^{m} w_i C_b(\mathbf{r}, \mathbf{r}_i)$$

$$+ \sum_{i=1}^{m} \sum_{j=1}^{m} w_i w_j [C_b(\mathbf{r}_i, \mathbf{r}_j) + C_o(\mathbf{r}_i, \mathbf{r}_j)], \quad (1.4)$$

where $E_b^2 = \langle e_b^2 \rangle$ is the mean-square background error, $C_b(\mathbf{r}_i, \mathbf{r}_j)$ and $C_o(\mathbf{r}_i, \mathbf{r}_j)$ are the covariance of the background and observational errors at locations $\mathbf{r}_i$ and $\mathbf{r}_j$, respectively, and $\mathbf{r}$ is the position of the analysis grid point. The minimum for (1.4) is found by setting the derivative with respect to weights w_i to zero. A set of linear algebraic equations are found that can be solved to give the optimum values for the weights w:

$$\sum_{j=1}^{m} w_j [C_b(\mathbf{r}_i, \mathbf{r}_j) + C_o(\mathbf{r}_i, \mathbf{r}_j)] = C_b(\mathbf{r}, \mathbf{r}_i). \quad (1.5)$$

With these optimum weights, the analysis error in (1.4) simplifies to

$$E_a^2 = E_b^2 - \sum_{i=1}^{m} w_i C_b(\mathbf{r}, \mathbf{r}_i) \qquad (1.6a)$$

or

$$E_a^2 = E_b^2 - \sum_{i=1}^{m} \sum_{j=1}^{m} w_i w_j [C_b(\mathbf{r}_i, \mathbf{r}_j) + C_o(\mathbf{r}_i, \mathbf{r}_j)].$$

$$(1.6b)$$

The solution of these equations requires a knowledge of the spatial structure of the covariance of the background forecast errors and of the observational errors, which can be obtained from a history of prior forecasts (see, e.g., Shaw et al. 1987). The geostrophic relation is then used to relate the various covariance functions between the mass and wind variables to that obtained for the mass field alone. The variable data density and accuracy is implicitly included in the minimization of the analysis error through the covariance between the observations. Since the scheme also gives an estimate of the analysis error, the difference between the observations and the interpolated value can be used to check for and remove suspect data (Lorenc 1981). An efficient scheme for the optimum interpolation of the mass field alone was devised by Bleck (1975), who utilized the wind observations as approximations to the geostrophic wind to extrapolate the mass observations to the grid points using (1.3) as in Cressman. Recently the method of optimum interpolation has been applied with success to operational limited-area modeling systems (DiMego 1988; Benjamin 1989; Mills and Seaman 1990). With limited-area forecast systems using optimal interpolation, the length scales of the correlation functions of the background forecast error are reduced from those used with global models to provide more detail at the smaller scales (DiMego 1988).

Bratseth (1986) introduced a successive correction scheme in which the solution converges toward the solution obtained by optimal interpolation. In his technique, the weights for each observation are given by the forecast error covariance, which is reduced in regions of higher data density by a factor given by the sum of the covariance between all the observations used in the interpolation. A further interpolation formula similar to (1.2) is also used to derive an estimate of the "observation value" itself, which converges to the actual value of the observation. This observation estimate is used for $\phi_{a,i}$ in (1.2) for the interpolation to the grid points. This is done instead of using an interpolated value derived by a polynomial approximation, as was done in earlier successive correction schemes. In this way, the interpolated values at the grid points converge to the values obtained by optimum interpolation and not to the data values. Since the Bratseth method requires much less computational expense than does optimal interpolation, it is very attractive for use for a limited-area forecast system when limited resources are available (Seaman 1988). This scheme is currently being used operationally in a multivariate analysis for the limited-area forecast system at the Norwegian Meteorological Institute (Gronas and Midtbo 1986). In their analysis, the rate of convergence of the estimate of the observation to the observation value itself is used for quality control of the data. The coupling of the geopotential to the geostrophic wind is reduced as the number of iterations increases, thereby limiting geostrophy to the larger scales. This becomes more important when used for higher-resolution models in which the mesoscale flow is more ageostrophic.

An algorithm similar to the Bratseth scheme has been applied by the U.K. Meteorological Office to insert data directly into both the global and regional models (Lorenc et al. 1991). With each iteration of the scheme, the mass and wind variables are updated in a sequential fashion. To maintain balance in the models, changes to the geostrophic component of the wind are derived from any mass changes, the wind changes are nondivergent, and the divergence during integration is damped. In scheme of Sashegyi et al. (1993), univariate analyses of the mass and wind fields are performed

using Bratseth's scheme. To provide the coupling between the wind and mass fields, they use the analyzed wind as an initial estimate of the geostrophic wind to extrapolate the geopotential to the gridpoint locations (as in Cressman 1959) for a further iteration on the geopotential. In subsequent iterations an improved geostrophic wind estimate is then defined by the new geopotential gradient, not by the original wind as was widely used with the Cressman scheme. The wind analysis is then corrected for the new geostrophic wind as in Kistler and McPherson (1975), and further iterations on the wind are then performed to improve the analysis of the larger scales of the ageostrophic component of the wind.

c. Variational adjustment with constraints

Sasaki (1958) introduced variational methods for the purpose of modifying analyses, so that they satisfy certain dynamic constraints such as the geostrophic condition, continuity equation, or the balance equation. The method was expanded by Sasaki (1969, 1970a,b) to include time variation and smoothing by low-pass filters in space or time. In the method, changes to the analysis are minimized in a least-squares sense, while at the same time the final result satisfies the constraints either locally at each grid point or in a global sense integrated over the domain.

We illustrate the method using the geostrophic condition as a constraint, following the notation in Daley (1991). For the changes to the wind field u_a, v_a, and geopotential field ϕ_a, a least-squares integral or cost function is defined by

$$J(u, v, \phi) = \frac{1}{2} \int\int [w_u(u - u_a)^2 + w_v(v - v_a)^2$$
$$+ w_\phi(\phi - \phi_a)^2]dxdy, \quad (1.7)$$

where the weights w_u, w_v, and w_ϕ are specified. A suitable choice for the weights are those defined by the inverse of the variance of the errors for each of the analyzed fields ($w_u = \sigma_u^{-2}$, $w_v = \sigma_v^{-2}$ for the wind and $w_\phi = \sigma_\phi^{-2}$ for the geopotential). In (1.7) we use the integral form for simplicity, while in the finite-difference form a summation is used. The cost function J is to be minimized subject to the following geostrophic constraints:

$$\frac{\partial\phi}{\partial x} - fv = 0, \quad \frac{\partial\phi}{\partial y} + fu = 0, \quad (1.8)$$

where f is the Coriolis parameter. The minimization is solved by taking the variation of J relative to the dynamical variables u, v, and ϕ and setting the result to zero, so that

$$\delta J = \int\int [w_u(u - u_a)\delta u + w_v(v - v_a)\delta v$$
$$+ w_\phi(\phi - \phi_a)\delta\phi]dxdy = 0. \quad (1.9)$$

This can be solved by the direct substitution of the constraints into (1.7), noting that the variations of u and v are related to the variations of ϕ by the same geostrophic expression as (1.8). Integrating by parts with $\delta\phi = 0$ on the boundary of the domain, we have

$$\delta J = \int\int \left\{ w_\phi(\phi - \phi_a) + \frac{\partial[w_u f^{-1}(u - u_a)]}{\partial y} - \frac{\partial[w_v f^{-1}(v - v_a)]}{\partial x} \right\} \delta\phi \, dxdy = 0, \quad (1.10)$$

where u and v are given by the constraints in (1.8). For the case $w_v = w_u$, the solution is then found by solving

$$w_\phi\phi - \frac{w_u}{f^2}\nabla^2\phi = w_\phi\phi_a - \frac{w_u}{f}\zeta_a \quad (1.11)$$

for the geopotential ϕ, where $\zeta_a = \partial v_a/\partial x - \partial u_a/\partial y$ is the analyzed vorticity, with suitable boundary conditions, such as $\phi = \phi_a$ on the boundary.

Alternatively, the minimum of the cost function J subject to the geostrophic constraints (1.8) can be found by the method of Lagrange multipliers. In this method a term containing the constraints is added to the cost function J:

$$J_1(u, v, \phi, \phi_x, \phi_y) = J + \frac{1}{2}\int\int [\lambda_1(x)(\phi_x - fv) + \lambda_2(y)(\phi_y + fu)]dxdy, \quad (1.12)$$

where ϕ_x and ϕ_y represent the partial derivatives of ϕ with respect to the coordinates x and y, respectively. Here, the Lagrange multipliers λ_1 and λ_2 are functions of x and y, respectively. Setting the variation of J_1 relative to the variables $u, v, \phi, \phi_x, \phi_y$ to zero, then (1.11) is again obtained. If the Lagrange multipliers used in (1.12) are instead constant values, then the constraints are satisfied only in a global sense integrated over the domain. The integral in this case is

$$J_1 = \frac{1}{2}\int\int [w_u(u - u_a)^2 + w_v(v - v_a)^2 + w_\phi(\phi - \phi_a)^2 + \lambda_1(\phi_x - fv) + \lambda_2(\phi_y + fu)]dxdy. \quad (1.13)$$

In the above examples, strong constraints are used, so that the solution satisfies the constraints exactly. As shown by Sasaki (1970a,b), the constraints can also be more loosely satisfied (weak constraints) by adding a penalty function to the cost function J, so that

$$J_2 = \frac{1}{2}\int\int [w_u(u - u_a)^2 + w_v(v - v_a)^2 + w_\phi(\phi - \phi_a)^2 + \gamma_1(\phi_x - fv)^2 + \gamma_2(\phi_y + fu)^2]dxdy, \quad (1.14)$$

where γ_1 and γ_2 are prescribed constants defining the degree of fit of the solution to the constraints. The smaller (larger) the values of γ, the weaker (closer) the fit of the final fields to the constraints.

The method has been applied by Stephens (1970) using a nonlinear balance equation as a constraint and by Achtemeier (1975) using the primitive equations as a constraint. As originally applied, the method was largely intended for the filtering of the high-frequency oscillations from the initial conditions for a numerical weather forecast model. In this regard, the method has been largely superseded by the normal-mode initialization scheme described in section 1.4. However, the variational method has direct applicability to the general analysis and assimilation problem itself, where analyzed or model fields are to be fitted in a least-squares sense to the observations in space and time (e.g., Ghil et al. 1981; Lorenc 1986). The close relationship between this method and optimum interpolation is illustrated for the generalized analysis in the next section.

d. Generalized analysis

In the statistical interpolation method described in section 1.3b the observations must be quantities that are the same as the dynamical variables of mass and wind in the model. Direct comparisons of model-derived quantities with measured values, such as radiances observed from satellites, cannot be used. The method of interpolation of the background estimates of mass and wind on a rectangular grid to the observation locations is also not considered. The statistical interpolation method can be generalized to address these problems (see, e.g., Ghil et al. 1981; Lorenc 1986). A generalized form of the linear interpolation algorithm for interpolating a set of observations $\mathbf{d}_o$ taken at one time to a rectangular grid can be written

$$\mathbf{s}_a = \mathbf{s}_b + \mathbf{K}(\mathbf{d}_o - \mathbf{H}\mathbf{s}_b), \quad (1.15)$$

where the elements of the vectors $\mathbf{s}_a$, $\mathbf{s}_b$, $\mathbf{d}_o$ represent the analyzed and background variables, and the value of each observation, respectively. Here, $\mathbf{K}$ is a gain matrix of weights to be determined, and $\mathbf{H}$ is a forward interpolation operator that represents the derivation of a first guess for the value of the observation from the background estimates. The errors can be written as

$$\mathbf{e}_a = \mathbf{e}_b + \mathbf{K}(\mathbf{e}_o + \mathbf{e}_f - \mathbf{H}\mathbf{e}_b) \quad (1.16a)$$

$$= (\mathbf{I} - \mathbf{KH})\mathbf{e}_b + \mathbf{K}(\mathbf{e}_o + \mathbf{e}_f), \quad (1.16b)$$

where the vectors $\mathbf{e}_a$, $\mathbf{e}_b$, $\mathbf{e}_o$, and $\mathbf{e}_f$ represent the errors of the analysis, background, observations, and forward interpolation operator $\mathbf{H}$, respectively. We assume that $\mathbf{H}$ is a linear operator and that the analysis, background, and interpolation errors are unbiased and not correlated with each other. Then the analysis error covariance matrix simply follows as

$$A = (I - KH)B(I - KH)^T + K(O + F)K^T, \quad (1.17)$$

where $A = \langle e_a e_a^T \rangle$, $B = \langle e_b e_b^T \rangle$, $O = \langle e_o e_o^T \rangle$, and $F = \langle e_f e_f^T \rangle$ are the error covariance matrices for the analysis, background, observations, and forward interpolation operator, respectively. Here, the transpose of a vector or matrix is represented by the superscript T. It can be noted that for the univariate case in (1.4), the elements of the covariance matrices B, O are just the covariances $C_b(r_i, r_j)$ and $C_o(r_i, r_j)$ of the background and observation errors for the variable ϕ, and the diagonal elements of A and B are the variances E_a^2 and E_b^2, respectively, at each of the grid points to be analyzed.

Now the mean-square analysis error $I = \langle e_a^T e_a \rangle$ is given by the sum of the diagonal elements of A (Ghil et al. 1981). The minimum of the mean-square analysis error is found by setting the derivative of I with respect to each of the elements in matrix K to zero. The result is

$$K = BH^T(O + F + HBH^T)^{-1} \quad (1.18)$$

for the optimum solution of K. For this optimum solution the analysis error covariance matrix simplifies to

$$A = (I - KH)B \quad (1.19a)$$

$$= B - BH^T(HBH^T + O + F)^{-1}HB. \quad (1.19b)$$

The generalization of (1.5) and (1.6a) to (1.18) and (1.19a) is readily apparent.

The result above can also be equivalently derived using the variational method described in section 1.3c for least-squares fitting (Lorenc 1986). The least-squares fit of the analysis to the data and the background in this case is given by a cost function J, which in finite-difference form is

$$J = \frac{1}{2}(d_o - Hs_a)^T(O + F)^{-1}(d_o - Hs_a)$$

$$+ \frac{1}{2}(s_a - s_b)^T B^{-1}(s_a - s_b), \quad (1.20)$$

where the weights are the inverses of the error covariance matrices. The minimum of the least-squares error function J is found by setting the derivative relative to each of the analyzed variables s_a to zero. In this case the analysis values found are the same as those defined by the minimum variance estimate given by (1.15) and (1.18), and the analysis error covariance matrix is also again given by (1.19b). If the background, observation, and interpolation errors are normally distributed, then this least-squares fit of J gives the most probable values for the analyzed variables s_a (Lorenc 1986), which are also the minimum variance estimates.

Such a generalized analysis is now the basis of a new statistical analysis scheme introduced in the operational global data assimilation system of the National Me-teorological Center (Parrish and Derber 1992). Currently, the scheme is too computationally expensive for use in many limited-area models. Further approximations to this generalized method will be required before such methods are of practical use for the analysis schemes of high-resolution limited-area models.

1.4. Initialization

Normal-mode initialization, which was developed by Machenhauer (1977) and Baer (1977), has been very successful in removing the high-frequency gravity wave oscillations from integrations of global numerical weather prediction models (see review by Daley 1981). In Machenhauer's scheme, the tendencies of the high-frequency gravity modes are set to zero at the start of the integration. To apply the scheme, it is first necessary to find the normal modes of the numerical model. The structure of the normal modes is solved by linearizing the equations of motion about a basic state at rest with a mean temperature that is a function only of the vertical coordinate used in the model. The horizontal pressure gradient force in the momentum equations for a general vertical coordinate is defined by both a pressure gradient term and a geopotential gradient term. When linearized, the two remaining linear terms are combined by defining a generalized geopotential whose horizontal gradient for a fixed value of the vertical coordinate gives the horizontal pressure gradient in Cartesian coordinates. The time tendency of this generalized geopotential is also related to the horizontal divergence in the model's coordinates (see, e.g., Temperton and Williamson 1981). A separation of the vertical and horizontal structure of the model variables is then possible. The resulting normal modes are eastward- and westward-propagating inertia–gravity modes, Kelvin modes, and Rossby modes (see, e.g., Daley 1991). By projecting the initial mass and wind fields onto these normal modes, the amplitudes of the high-frequency inertia–gravity modes can be adjusted to match the nonlinear forcing of these modes, thereby satisfying Machenhauer's condition.

For each of the normal modes, the equations of motion can be written as

$$\frac{dc_k}{dt} = -i\omega_k c_k + r(c_l, c_k), \quad (1.21)$$

where c_k is the complex coefficient of the normal mode and ω_k is its frequency. The nonlinear terms are represented by the last term, r, which depends on all the modes. In Machenhauer's scheme, the time derivative for the high-frequency modes is set to zero at the initial time to give

$$c_k = \frac{r(c_k, c_l)}{i\omega_k}, \quad (1.22)$$

which must then be solved iteratively. As it is usually applied in numerical forecast models, the required change to each of the initial complex coefficients c_k of the "fast" gravity modes is given by the tendency for each fast mode, computed by integrating the model forward one time step, where

$$c_k(n + 1) = c_k(n) + \frac{1}{i\omega_k} \frac{dc_k(n)}{dt}, \quad (1.23)$$

where ω_k is the frequency for that fast inertia–gravity mode. Several methods currently being used to apply this scheme to limited-area models are now described.

a. Vertical-mode initialization

Briere (1982) applied nonlinear normal-mode initialization to a limited-area model, computing the normal modes by linearizing the equations of motion with constant Coriolis parameter (i.e., for an f-plane linearization); constant map factors; and, where the values of the model variables were assumed constant, at the lateral boundaries. When the boundary conditions vary with time or variable map factors are used, it is difficult if not impossible to solve for the horizontal structure of the modes. However, in this case the vertical structure of the normal modes can still be computed for the linearized equations of motion. The shallow-water equations result from projecting the mass and wind fields onto these vertical "modes" to remove the dependency on the vertical structure in the equations of motion. Bourke and McGregor (1983) applied filtering conditions to remove the inertia–gravity waves from the resulting shallow-water equations and showed that they were related to nonlinear normal-mode initialization when using an f-plane linearization. This equivalence to nonlinear normal-mode initialization is further expanded upon by Juvanon du Vachat (1986) and Temperton (1988). The approach is also extended by Temperton (1988) to include the β term in the linearization and to use higher-order time derivatives for the filtering conditions. Sashegyi and Madala (1993) have applied the vertical-mode initialization scheme of Bourke and McGregor (1983) to a limited-area weather prediction model formulated in flux form, using an alternative approach to the boundary conditions in the scheme. In the vertical-mode schemes, the filtering conditions are applied for only those vertical modes of the model for which the phase speeds of the freely propagating gravity modes are larger than the typical speeds of weather systems that are less than 25 m s^{-1}. A frequency cutoff, which is applied to each individual normal mode in the normal-mode initialization, cannot be used in this case.

In the vertical-mode initialization schemes above, the vertical structure of the normal modes in the equations of motion is extracted, and the three-dimensional flow can be represented by the shallow-water equations for each vertical mode. The resulting equations are expressed as time tendencies of vorticity ζ, divergence

D, and generalized geopotential Φ for each vertical mode by

$$\frac{\partial \zeta}{\partial t} + fD = A_\zeta, \quad (1.24)$$

$$\frac{\partial D}{\partial t} + \nabla^2 \Phi - f\zeta = A_D, \quad (1.25)$$

$$\frac{\partial \Phi}{\partial t} + gh_k D = A_\Phi. \quad (1.26)$$

Here, h_k is the equivalent depth for the kth vertical mode; f is the Coriolis parameter, which can be allowed to vary with latitude; and g is the acceleration due to gravity. The nonlinear terms A_ζ, A_D, A_Φ on the right-hand side (rhs) of the equations include also the diabatic and frictional forcing. The gradient β of the Coriolis parameter is also treated as a nonlinear term and included on the rhs in the equations.

For the f-plane linearization used, the solutions to the homogeneous shallow-water equations [(1.24), (1.25), and (1.26) with the forcing terms on the rhs equal to zero] are freely propagating inertia–gravity waves. The amplitude of each gravity mode depends only on the projection of the divergence D and the ageostrophic "vorticity" ($f\zeta - \nabla^2\Phi$) onto the horizontal structure of that normal mode (Leith 1980). The frequency of the free inertia–gravity modes is given by $(f^2 + k^2 gh_k)^{1/2}$, where k is the wavenumber. A further property of these homogeneous equations is that the linearized potential vorticity ($\zeta - f\Phi/gh_k$) is unchanged by the motion of the freely propagating inertia–gravity modes (see, e.g., Temperton 1988). The first vertical mode with the largest equivalent depth is the external (barotropic) mode with nearly constant amplitude in the vertical direction and a maximum phase speed of about 300 m s^{-1}. With decreasing equivalent depth, the vertical structure of the mode has more nodes (zero crossings) in the vertical ($k - 1$ nodes for the kth vertical mode), and the frequency and phase speed of the vertical mode decreases.

The filtering conditions of Bourke and McGregor require that, initially, the time tendencies of the divergence and the ageostrophic vorticity are zero and that the linearized potential vorticity does not change for each of the first three vertical modes with the largest equivalent depths. Not surprisingly for the f-plane linearization, the scheme is the equivalent in physical space to applying Machenhauer's condition on the amplitude of the fast gravity modes (Bourke and McGregor 1983; Juvanon du Vachat 1986; Temperton 1988). When applied to (1.24), (1.25), and (1.26) with the forcing on the rhs included, the freely propagating inertia–gravity waves are eliminated. Since the terms on the rhs of (1.24), (1.25), and (1.26) depend on the vorticity, divergence, and geopotential, respectively, the resulting set of equations is solved iteratively for each of the first three vertical modes (Bourke and

McGregor 1983). After some manipulation the resulting equations are usually written in terms of the changes in the geopotential, divergence, and vorticity— $\Delta\Phi$, ΔD, and $\Delta\zeta$, respectively—at each iteration as

$$\left(\nabla^2 - \frac{f^2}{gh_k}\right)\Delta\Phi(n + 1) = \frac{\partial D(n)}{\partial t}, \quad (1.27)$$

$$\left(\nabla^2 - \frac{f^2}{gh_k}\right)\Delta D(n + 1)$$

$$= -\frac{1}{gh_k}\frac{\partial}{\partial t}[f\zeta(n) - \nabla^2\Phi(n)], \quad (1.28)$$

$$\Delta\zeta(n + 1) = \frac{f\Delta\Phi(n + 1)}{gh_k}, \quad (1.29)$$

where the terms on the rhs of (1.27) and (1.28) are residuals given by the tendencies of the divergence and the ageostrophic vorticity. These tendencies are computed at each iteration by integrating the model one time step usually without friction, diffusion, or updating of the values of the model variables at the lateral boundaries. The usual boundary conditions are no change in the generalized geopotential and the mass divergence at the lateral boundaries of the model domain. Temperton (1988) shows that the above set of equations can be rewritten in the same form as (1.23), to explicitly relate the changes in the mass, divergence, and vorticity for the vertical modes to the tendencies of mass, divergence, and vorticity, respectively, for the gravity modes alone.

b. Dynamic initialization

The time integration scheme in the numerical model can also be used to remove the high-frequency modes (Miyakoda and Moyer 1968; Nitta and Hovermale 1969). In this dynamic initialization procedure, a forward and backward integration of the numerical model is carried out about the initial time, followed by a linear combination of the initial and final states. By repeated iterations, the high-frequency modes are selectively damped. More efficient schemes were developed by Sugi (1986) and Bratseth (1982), in which the nonlinear terms were held constant during each forward and backward integration. Their schemes can be shown to be equivalent to Machenhauer's scheme used for normal-mode initialization. Applying a forward and backward integration to the normal mode (1.21), we have

$$c^*(n + 1) = c(n) - i\omega\Delta t c(n) + r(n)\Delta t, \quad (1.30a)$$

$$c^{**}(n) = c^*(n + 1) + i\omega\Delta t c^*(n + 1) - r(n)\Delta t, \quad (1.30b)$$

where n is the iteration number and Δt is the time step. The new estimate is then given by the linear combination

$$c(n + 1) = c(n) - \gamma[c^{**}(n) - c(n)], \quad (1.31)$$

where γ is a factor to be chosen. This then gives

$$c(n + 1) = (1 - \gamma\omega^2\Delta t^2)c(n) - i\gamma\omega\Delta t^2 r(n) \quad (1.32)$$

or

$$c(n + 1) = c(n) + \frac{\gamma\omega^2\Delta t^2}{i\omega}\frac{dc(n)}{dt}. \quad (1.33)$$

If the scheme converges [i.e., $c(n + 1) = c(n)$ for some large n], then we have

$$\frac{dc(n)}{dt} = 0, \quad (1.34)$$

which is the same as Machenhauers condition. In fact Bratseth showed that by choosing $\gamma = (\omega\Delta t)^{-2}$ for each high-frequency mode and $\gamma = 0$ for the slow meteorological modes, the scheme is exactly the same as Machenhauer's scheme. When the individual modes cannot be calculated, a maximum frequency ω_m can be computed for each vertical mode using the maximum phase speed of $(gh)^{1/2}$ for that vertical mode and the smallest resolvable wavelength (twice the grid spacing in the model), and then $\gamma = (\omega_m\Delta t)^{-2}$. In this case the iterative scheme is an underrelaxation modification similar to that used by Kitade (1983) for normal-mode initialization.

c. Laplace transform method

In the Laplace transform technique of Lynch (1985a,b), the model equations are transformed into the complex plane by the Laplace transform. Wave motions of frequency ω are transferred into rational functions of a complex variable s with denominator ($s - i\omega$). The high-frequency components then correspond to poles at the points $s = i\omega$ in the complex plane far from the origin. The inverse transform back to physical space is carried out by a contour integral in the complex plane. The high-frequency components can be simply eliminated by the selection of a closed contour of integration around the origin that does not include the poles of the high-frequency components. The equivalence with nonlinear normal-mode initialization can be demonstrated by taking the Laplace transform L of (1.21) for the normal modes:

$$(s - i\omega)\hat{c}(s) = \frac{r(0)}{s} + c(0), \quad (1.35)$$

where $\hat{c}(s) = L(c)$, and the nonlinear terms r are treated as constant. Equation (1.35) can be rewritten as

$$\hat{c} = \left[c(0) + \frac{r(0)}{i\omega}\right]\frac{1}{s - i\omega} - \frac{r(0)}{i\omega s}. \quad (1.36)$$

On applying the inverse transform around a contour circling the origin but not enclosing the pole at $s = i\omega$, the first term does not contribute to the integral and we obtain

$$c = \frac{r}{i\omega}, \qquad (1.37)$$

which is just Machenhauer's scheme. The method does not require computing the structure of the normal modes and can be applied directly to the equations for limited-area models (Lynch 1985b).

d. Limitations of normal-mode initialization

The normal-mode initialization methods described in this section were applications of Machenhauer's scheme to limited-area models. In Machenhauer's scheme, it is required that the inertia–gravity modes are of a much higher frequency than the nonlinear forcing by the meteorologically significant slower modes. This is the case for gravity modes of large equivalent depth h_k (low vertical-mode number), and in this case Machenhauer's scheme is very successful in filtering out the high-frequency gravity waves. However, as the equivalent depth is reduced (for higher mode number), or the wavelength is substantially increased, the frequency of the inertia–gravity modes is decreased. The time tendency term cannot then be ignored if the horizontal scale of the mode is much larger than the Rossby radius of deformation $(gh_k)^{1/2}/f$, and Machenhauer's condition is inappropriate. For this reason, the schemes described in this section should be applied to the first two or three vertical modes only.

The fact that these first three vertical modes are largely balanced has been demonstrated for simulations in a global model by Errico and Williamson (1988) and in a mesoscale model by Errico (1990). For the shallower modes, however, they showed that the local time tendencies were large compared to the nonlinear forcing. It is not surprising then that the normal-mode initialization scheme has been successful in producing realistic ageostrophic flows in the regions of such deep mesoscale circulations as jet streaks in the upper troposphere (Sashegyi et al. 1993). For the shallower mesoscale circulations with a large ageostrophic component to the flow, a period of assimilation during which time the model can spin up such circulations is required. Such assimilation methods, which can be used to generate initial conditions for model forecasts, are now discussed.

1.5. Assimilation

Due to the incomplete data coverage, methods of incorporating data at earlier times are required to produce the most complete information possible on the current state of the atmosphere. Although the new observing systems—such as NEXRAD (Next Generation Weather Radar) Doppler radars, acoustic sounders, and higher-resolution dropwindsondes—will provide a higher frequency of observations, they will cover smaller areas. In all these methods of incorporating the data, the numerical model is integrated in time to provide the background forecast field for the introduction of data corrections. By using a mesoscale model to generate a first guess, realistic mesoscale features can be generated by the model, which will be preserved or even enhanced with the introduction of the data, especially from these newer observing systems. In this section we discuss the current methods commonly used for assimilating data in a limited-area model and introduce briefly the newer Kalman–Bucy filtering and variational adjoint methods that are being developed. Further details can be found in Daley (1991) and the references cited therein. These assimilation methods can be generally classified as either continuous methods or intermittent methods of data assimilation.

a. Intermittent data assimilation

The most common method of intermittent data assimilation used in operational weather centers is the analysis–forecast update cycle. In this approach, a forecast from a previous run of the numerical model is used as the background for an analysis of the current data. All the observations taken with a fixed time period, such as 3 h, centered around the analysis time are collected and used in the analysis. A nonlinear normal-mode initialization is usually then carried out to remove noise. The new initialized fields are used to generate a new prediction that is then used as the background for the data at the next analysis time, and the cycle is repeated. The length of the cycles typically varies from 3 to 12 h. For limited-area models that are not run in such an operational mode, an initial background forecast can be generated from an earlier larger-scale analysis to start one or more update cycles for a period of 12–24 h of assimilation. Rogers et al. (1990) found only a small impact on an operational regional assimilation system when extra synoptic soundings taken during the Genesis of Atlantic Lows Experiment were included in the update cycle every 12 h. Harms et al. (1992b) assimilated the observed precipitation and the supplemental 3-h synoptic and dropwindsonde soundings in a higher-resolution limited-area model using an update cycle. In their case the assimilation of these data produced stronger thermal and moisture gradients in the boundary layer and much better precipitation in the first 12 h of the subsequent forecasts.

b. Continuous data assimilation

For data that are not observed at such regularly spaced intervals or are more frequent than a couple of hours, a method of continuous assimilation is more appropriate. In these methods, updating of the model by the data is carried out at each time step of the model integration during a fixed time period of assimilation.

The method of nudging is commonly used, which adds a Newtonian relaxation term to the equations of motion to drive the predicted fields to the data (Hoke and Anthes 1976; Davies and Turner 1977). For example, for the u component of the wind field, the equation of motion is modified by

$$\frac{\partial u}{\partial t} = F + G(r, t)(u_0 - u), \qquad (1.38)$$

where F represents the nonlinear and forcing terms in the equation of motion. In the additional nudging term, u_0 can represent an individual observation or a value interpolated in time from analyzed fields. The function $G(r, t)$ is a specified weighting function that spreads the influence of the data in time and space. The time variation of the weighting function is usually chosen to slowly increase (decrease) prior to (after) the time of the observation to prevent any shocks to the model during the assimilation process. To further control noise, an extra divergence damping term is often also included (Lorenc et al. 1991). Nudging has been successfully tested for use with the new generation of data-observing systems—such as dropwindsondes by Ramamurthy and Carr (1987), the wind profilers by Kuo and Guo (1989), and surface data by Stauffer et al. (1991). In all these applications, a disadvantage is that the weights are not based on a statistical least-squares fit.

The generalized analysis and variational methods described in section 1.4 can also be extended to include the variation with time by utilizing the numerical model. Following Daley's (1991) notation, we can represent a single time step of the numerical model by

$$\mathbf{s}_f(n + 1) = \mathbf{M}\mathbf{s}_f(n), \qquad (1.39)$$

where the vector $\mathbf{s}_f(n)$ represents the model variables at a single time, and matrix $\mathbf{M}$ represents the effect of integrating the model forward by one time step. In the variational method, the assimilated fields are integrated over a period of time and the initial conditions are successively adjusted to minimize the error. The error is measured by the least-squares fit of the model assimilated fields to the data as described by the cost function J in (1.20). Summing the first term in (1.20) for each time step during the period of assimilation, we have

$$J = \frac{1}{2} \sum_{n=0}^{N} [\mathbf{d}_o(n) - \mathbf{H}\mathbf{s}_f(n)]^\mathrm{T} \mathbf{R}^{-1}(n)[\mathbf{d}_o(n) - \mathbf{H}\mathbf{s}_f(n)],$$

$$(1.40)$$

where $\mathbf{R}$ is the observation error covariance matrix and N is the number of time steps. The error due to the model background forecast [second term in (1.20)] is usually ignored. To compute the minimum of J, the gradient of J relative to changes in the model variables is computed for each time step by a backward integration of the adjoint equation for the model (Tala-

grand and Courtier 1987; Courtier and Talagrand 1987). The initial conditions are then modified by applying a descent algorithm using the gradient of J computed for the initial time. The forward integration of the model and the backward integration of the adjoint model are then repeated to successively adjust the initial conditions and minimize the error. Noise can be controlled by the addition of a penalty term to the cost function J as in (1.14). Much work still needs to be done before the method will be practical for data assimilation in mesoscale models.

In the Kalman–Bucy filtering technique (Ghil et al. 1981), the data sequentially adjust the assimilated fields as the model is integrated forward in time. When data are available, an optimum analysis—such as the generalized analysis scheme described by (1.15), (1.18), and (1.19) in section 1.3d—is used to give best current analysis with the model forecast providing the background field. The model is then used to integrate forward one time step to produce a new forecast field

$$\mathbf{s}_f(n + 1) = \mathbf{M}\mathbf{s}_a(n), \qquad (1.41)$$

where the vector $\mathbf{s}_a(n)$ represents the analyzed model variables at a single time, and matrix $\mathbf{M}$ represents the effect of integrating the model forward by one time step to produce a forecast $\mathbf{s}_f(n + 1)$. If the model matrix $\mathbf{M}$ is assumed to be linear, then the forecast errors $\mathbf{e}_f$ can be computed according to

$$\mathbf{e}_f(n + 1) = \mathbf{M}\mathbf{e}_a(n) - \mathbf{e}_q(n), \qquad (1.42)$$

where the vectors $\mathbf{e}_a$, $\mathbf{e}_q$ represent the analysis errors and the errors in the model itself. The forecast error covariance matrix $\mathbf{P}_f(n + 1)$ then follows

$$\mathbf{P}_f(n + 1) = \mathbf{M}\mathbf{P}_a(n)\mathbf{M}^\mathrm{T} + \mathbf{Q}(n), \qquad (1.43)$$

where $\mathbf{P}_a(n)$ is the analysis error covariance matrix and $\mathbf{Q}(n)$ is the model error covariance matrix. The new forecast error covariance can then be used in the next optimum analysis. To control noise, initialization can be included by a modification to the weights matrix $\mathbf{K}$ used in (1.15) by a projection of the analysis increments onto the slow Rossby modes only. Currently, however, this method is very expensive to implement.

c. Assimilation in the planetary boundary layer

With high-resolution mesoscale models, use of the mesoscale model to generate an initial guess of the vertical structure of the moisture and the flow in the planetary boundary layer is essential. Data from the synoptic network or from an analysis from a regional modeling system do not provide the required detail for a high-resolution mesoscale model. In particular these data are not representative of the flow over regions of highly variable terrain and surface heating in the mesoscale model. In the past, one-dimensional boundary layer models have been used to correct the low-level winds in the synoptic soundings, which were used to provide the initial conditions for the mesoscale model (e.g., Segal and Pielke 1981). Since in midlatitudes the temperature in the boundary layer responds more quickly than the winds at the small scales, only the low-level winds needed to be adjusted in their case (see Pielke 1984, p. 357). Golding (1987) and Ballard et al. (1991) used a surface analysis of humidity and the observed cloud type to manually adjust the initial humidity fields, which had been interpolated from a regional model. The initial humidity and cloud distribution in the mesoscale model was much better represented and the precipitation forecast was more accurate in the first 3 h. Lipton and Vonder Haar (1990a,b) assimilated satellite retrievals of water vapor and surface ground temperatures in a mesoscale model successfully using a forecast analysis update cycle. Such use of retrieval algorithms and other parameterizations for the boundary layer processes will be needed for utilizing satellite data in the future in combination with mesoscale models.

1.6. Lateral boundary conditions

Specification of accurate boundary conditions is very important in high-resolution limited-area models. Small-scale errors in the initial conditions can be removed by the advection of information into the domain from the lateral boundaries in regions of inflow (Errico and Baumhefner 1987). Also, if the scale of the evolving disturbance is large compared to the domain of the limited-area model, the lateral boundary conditions act to constrain the solution, thereby further reducing the error (Vukicevic and Errico 1990). Boundary conditions for limited-area models are usually provided by a forecast from a global model, for example. Conversely, any serious errors in the lateral boundary conditions obtained from such larger-scale forecasts can quickly contaminate the forecast produced by a limited-area model. This is particularly a problem in rapidly evolving synoptic situations, such as occurs with the passage of a front through the domain. The forecasts produced by the limited-area model are then limited in accuracy by the accuracy of the boundary conditions obtained from the global forecast model (Vukicevic and Errico 1990). However, strong forcing of the flow by topography or surface heating can also result in more predictable solutions on the mesoscale in weak synoptic regimes. These aspects relating to such surface forcing can be found discussed in Pielke (1984). Here we will limit ourselves to a discussion of the commonly used methods for specifying the lateral boundary conditions in limited-area models.

In a fully three-dimensional flow, the implementation of mathematically correct lateral boundary conditions is difficult (see, e.g., Sundstrom and Elvius 1979). Several pragmatic approaches are therefore used to formulate the boundary conditions in limited-area models. The most common are based on the tendency damping scheme of Perkey and Kreitzberg (1976), the Newtonian relaxation scheme of Davies (1976), and the Sommerfeld radiation condition of Orlanski (1976). In the Perkey and Kreitzberg (PK) scheme, model-computed tendencies are damped at each time step to specified boundary tendencies in a boundary zone. In the Davies scheme, computed model variables themselves are relaxed at each time step to the boundary values in a boundary zone. The width of the boundary zone is usually five to six points wide and the coefficients used to linearly combine the boundary values (tendencies) with the interior solution values decrease with distance from the boundary, from one at the boundary to zero in the interior. In the various adaptations of the Orlanski scheme, the phase speed (which can include an advection velocity) at outflow points along the lateral boundary is estimated from the values of the solution at interior points (Miller and Thorpe 1981).

For a limited-area model, the lateral boundary formulation can act to damp the errors in the solution in the region of the boundary and thereby in time remove those in the interior as they propagate away into the boundary region. These errors can also include gravity wave oscillations produced by unbalanced initial conditions in the interior of the model domain. In the boundary zone, the PK scheme reduces the advection speed of the solution error, with the advection speed decreasing to zero at the boundary (Davies 1983). The wavelength of the solution error is also reduced as the errors propagate toward the lateral boundary because of the progressively smaller weighting function applied to the internal tendency terms (while weights on the values from the boundary solution become greater). The phase speed and wavelength of the gravity waves is similarly reduced in the boundary zone. The result is that these errors can be removed as they slow in the boundary zone by application of increased diffusion near the boundary. The phase speed of waves entering the domain are, however, also retarded as they pass through the boundary zone. With the Davies scheme, the amplitude of the errors in the solution (including the gravity waves) are damped in the boundary zone, with the damping increasing closer to the boundary (Davies 1983). However, while interior errors propagating into the boundary region are damped in this case, errors in the boundary values are free to propagate into the interior. When synoptic forcing through the boundaries is important, the Davies scheme is usually the preferred scheme for the lateral boundary conditions.

Boundary conditions derived from global or regional forecasts can also produce errors when they are too coarse in space and time for the finer resolution mesoscale model in which they are used. These errors may be reduced by using successive nested grids of less extreme resolution change to reach the target mesoscale model. Also, the use of a higher-order interpolation scheme can help to preserve features in the solutions from which the boundary values are derived, such as frontal shapes (e.g., Williamson and Rasch 1989).

1.7. Conclusions

The importance of using the same mesoscale model for the generation of the initial data as will be used later for the prediction or simulation of the mesoscale weather event has been emphasized. Global- or regional-scale analyses are of too coarse a resolution to initialize a mesoscale model and often a single sounding is not representative of the flow over the variable terrain and surface conditions in the model domain. Use of the mesoscale model to generate a background forecast for a data analysis and/or using a period of assimilation for introducing the data can greatly enhance the resolution of the initial conditions. The nesting of the mesoscale model inside a high-resolution limited-area model is helpful to assimilate the data on the synoptic scale and thereby provide the boundary conditions with the temporal and spatial resolution required by the mesoscale model. The future increase in remotely sensed data available from satellites and radars will necessitate further research in using various retrieval algorithms and reverse schemes to adjust the initial conditions to match the observed quantities (such as radiances and rainfall) and to be consistent with the model physical parameterizations in the boundary layer. The higher frequency of observations that will be available from many of the new observing systems, such as acoustic sounders and Doppler radars, will place severe demands on current assimilation methods. The future application of methods of assimilation based on the newer adjoint method and Kalman–Bucy filtering may hold promise for use with such high-frequency data in mesoscale models.

Acknowledgments. This research was sponsored by NRL's basic research program and by SPAWAR of the United States Navy. This paper was presented at the Mesoscale Modelling Workshop Incorporating Project WIND Data, 16–18 June 1992, El Paso, Texas. The workshop and Project WIND were sponsored by the United States Army Atmospheric Sciences Laboratory, White Sands Missile Range, New Mexico.

Chapter 2

Subgrid-Scale Parameterizations

MAREK ULIASZ

Department of Atmospheric Science, Colorado State University, Fort Collins, Colorado

2.1. Introduction

Mesoscale meteorological models are governed by nonlinear equations of motion, and continuity equations for mass, heat, and water substance. These equations contain information on atmospheric motion and transport over different scales down to the smallest eddies responsible for the viscous dissipation. Since it is not possible to integrate exactly such a set of equations, the modeler has to distinguish between those eddies that can be resolved by a numerical model and the eddies that are not fully resolved computationally and, therefore, are defined as a subgrid-scale process or turbulence. On the other hand, one might not want detailed information about eddies with sizes that cannot be resolved by an available observational system.

The general approach is to decompose each model variable into a mean and a turbulent fluctuating component, $\phi = \bar{\phi} + \phi'$; average the model equations; and solve for averaged variables. Averaging operators used in mesoscale and clouds models and their properties and implications in forecasting are discussed in depth by Cotton (1986). In this chapter ensemble averaging is discussed, which represents the most likely value of a subgrid-scale quantity, whereas grid volume averaging represents just one realization. If the Boussinesq set of equations in anelastic approximation for a dry atmosphere is considered as an example, these equations after averaging become

$$\frac{\partial \overline{u_i}}{\partial t} = -\frac{\partial (\overline{u_i u_j})}{\partial x_j} - \frac{\partial (\overline{u_i' u_j'})}{\partial x_j}$$

$$-\frac{1}{\rho_0}\frac{\partial \bar{P}}{\partial x_i} - 2\varepsilon_{ijk}\Omega_j \overline{u_k} + \delta_i \beta \bar{\vartheta} \quad (2.1)$$

$$\frac{\partial \bar{\vartheta}}{\partial t} = -\frac{\partial (\overline{u_j}\,\bar{\vartheta})}{\partial x_j} - \frac{\partial (\overline{u_j' \vartheta'})}{\partial x_j} + \frac{Q_R}{\rho_0 C_p} \quad (2.2)$$

$$\frac{\partial \overline{u_i}}{\partial x_i} = 0 \quad (2.3)$$

$$\bar{P} = \rho_0 R \bar{T}, \quad (2.4)$$

where u_i, $i = 1, 2, 3$ are wind velocity components, ϑ is potential temperature, P is pressure, Q_R is the radia-

tive heat source, $\beta = g/\vartheta_0$, and conventional meteorological notation is used for other variables. The main difference between the original and averaged set of equations is the appearance of the second-order moments: subgrid-scale momentum fluxes $\overline{u_i' u_j'}$ and subgrid-scale heat fluxes $\overline{u_i' \vartheta'}$. It is possible by manipulation of the model equations to derive prognostic equations for the second-order moments. Unfortunately, these equations contain in turn unknown third-order moments. The equations for third-order moments contain fourth-order moments and so on to higher order. The number of unknowns in the set of equations for turbulent flow is always larger than the number of equations, so these equations are not closed. A variable is considered unknown if there is no prognostic or diagnostic equation to define it. To close the set of equations, the unknown variables must be evaluated through the known variables. This specification of subgrid-scale processes using experimental data and simplified fundamental principles is called *parameterization*. Usually parameterizations are not defined in terms of basic conservation principles. Some parameterization rules for turbulent closures are discussed, for example, by Stull (1988). Closure schemes are named by the highest-order prognostic equations that are retained. The *first-order closure* models retain only prognostic equations for mean variables, and turbulent fluxes (second-order moments) are expressed by the mean variables at a particular level. The *second-order closure* models employ equations for the second moments that are closed on the level of third moments. In general, *n*th-order closure involves equations for the *n*th moments closed on the level of $n + 1$ moments. Some closure assumptions utilize only a portion of the equations available within a particular moment category.

The turbulent closure techniques can be also classified as *local* and *nonlocal* closures. For local closure, an unknown variable at any point in space is parameterized by values or gradients of known variables at the same point. Nonlocal closure assumes that the turbulence is a superposition of eddies transporting fluid like an advection process; therefore the unknown variable depends on fluid properties in many points.

The averaging may introduce in addition to turbulent fluxes other types of subgrid-scale processes that must be parameterized (Pielke 1984): averaged radiation flux divergence and averaged effect of the change of phase of water including precipitation. However, this chapter is limited to discussion of the parameterizations of the averaged vertical turbulent fluxes in mesoscale models. Horizontal subgrid-scale fluxes are utilized mainly for computational reasons since little is known about horizontal subgrid-scale mixing on mesoscale.

Many parameterizations implemented in mesoscale models were originally developed for one-dimensional atmospheric boundary layer (ABL) models where different approaches are used for the turbulence closure problem. It should be noted that after considerable effort in boundary layer research, a generally accepted ABL theory has not evolved like, for example, Monin–Obukhov similarity theory for the surface layer (Hasse 1993). On the other hand, turbulence parameterizations for mesoscale models may cause some additional problems. Mesoscale modeling domains extend in the vertical usually above the ABL, often covering the whole troposphere. Special attention must be paid to turbulence parameterization in the free atmosphere (e.g., related to orographical effects) as well to parameterizations for mesoscale circulations in complex terrain, which often result in internal boundary layers. In general, these parameterizations have to deal with a multilayer structure of turbulence in the atmosphere, while the vertical resolution of mesoscale models is typically much lower than in the ABL models.

This review starts from the simplest first-order parameterizations. Next, higher-order closures are discussed with emphasis on turbulent kinetic energy (TKE) closure, which is perhaps the most popular closure scheme used currently in mesoscale models. Discussion of length-scale formulations and some aspects of nonlocal closures are then presented. Finally, turbulence parameterizations for the atmospheric surface layer are shortly reviewed.

Since this review is limited to the models that can resolve the ABL, integral (slab or bulk) parameterizations are not discussed here. This approach, which can be generally classified as a *half-order closure,* reduces the closure problem to the assumptions concerning mean profiles of wind and temperature within the ABL, surface turbulent fluxes, and evolution of the ABL height. The bulk parameterizations are used to represent boundary layer processes in large-scale and global circulation models (e.g., Deardorff 1972; Stull 1973; Brost and Wyngaard 1978; Randall and Shao 1992).

2.2. First-order closure

The first-order closure called gradient transport or K theory uses a formal analogy between molecular diffusion and turbulent diffusion. The vertical turbulent fluxes are expressed with the aid of local gradients of corresponding mean variables:

$$\overline{w'\phi'} = -K_\phi \frac{\partial \overline{\phi}}{\partial z}, \qquad (2.5)$$

where K_ϕ is called eddy diffusivity or turbulent exchange coefficient and ϕ represents any model variable. The eddy diffusivity for heat, $K_h = \alpha_T K_m$, is often assumed to be proportional to the eddy diffusivity for momentum, K_m, and the diffusivities for heat and water vapor are assumed to be equal. Unlike molecular diffusion, turbulent diffusion is not a function of the fluid but is a function of the flow. Therefore, eddy diffusivities can depend on time and space but are supposed to remain positive. This implies that the turbulent flux $\overline{w'\phi'}$ is directed down the gradient of the variable $\overline{\phi}$, which is a serious restriction to the generality of this formulation since countergradient turbulent fluxes are often observed in the atmosphere (e.g., Deardorff 1966). Use of K theory is not recommended in convective mixed layers where large eddies associated with rise of warm air parcels can transport heat regardless of the local gradients in the background atmosphere.

Despite its limitations, K theory has been shown to be a useful parameterization of turbulent fluxes in meteorological models. A large variety of formulations for eddy diffusivities can be found in literature (e.g., Blackadar 1979; Bodin 1980; Sorbjan and Uliasz 1982). From a methodical point of view, all K parameterizations can be classified into two categories: 1) empirical formulations based on experimental data, and 2) semiempirical formulations that can be derived from prognostic equations for second moments. Within the empirical parameterizations, eddy diffusivities can be prescribed explicitly as a function of height or implicitly as a function of the mean variables and their gradients at a given level.

Explicit parameterizations usually utilize an eddy diffusivity profile derived from the surface-layer similarity theory [see (2.27) in section 2.6] close to the ground surface and a certain interpolation formula within the ABL. An example is the well-known O'Brien profile (O'Brien 1970), where K_ϕ is approximated by a cubic polynomial with given K_ϕ and dK_ϕ/dz at the surface-layer top h_s from the similarity theory and a small value $K_{\min}$ at the top of the boundary layer h_{ABL}. O'Brien's profile was applied quite often in mesoscale modeling both with a fixed h_{ABL} height and in combination with h_{ABL} prediction (e.g., Bornstein 1975; Pielke and Mahrer 1975).

The explicit K formulations are economic and can provide satisfactory results when applied in situations adequate to physical assumptions on which they are grounded. However, their application is limited to the ABL, whereas the computational domain in mesoscale models often includes the whole troposphere. Their main disadvantage is the fixed shape of the K profile, which does not allow correct simulation of internal boundary layers. These restrictions are avoided in the implicit formulations that can be presented in a general form as functions of Richardson number Ri, mixing length scale l_k, and wind shear S:

$$K_\phi = l_k^2 S f(\text{Ri}),\qquad(2.6)$$

where

$$S = \left[\left(\frac{\partial \bar{u}}{\partial z}\right)^2 + \left(\frac{\partial \bar{v}}{\partial z}\right)^2\right]^{1/2},\quad \text{Ri} = \frac{\beta}{S}\frac{\partial \bar{\vartheta}}{\partial z}.\quad(2.7)$$

The formulations for the function $f(\text{Ri})$ are numerous (e.g., Blackadar 1979). This function is sometimes specified on a purely empirical basis, but it can also be derived from a simplified second-order closure (Mellor and Yamada 1974) or from asymptotic arguments as in Louis (1979):

$$f(\text{Ri})$$
$$=\begin{cases} 1 - \dfrac{b\,\text{Ri}}{1 + c|\text{Ri}|^{1/2}}, & \text{unstable stratification} \\[3mm] 1 + \dfrac{1}{(1 + b'\,\text{Ri})^2}, & \text{stable stratification,} \end{cases}$$
$$(2.8)$$

where b, b', and c are nondimensional constants. It should be noted that dependence of eddy diffusivity on Richardson number is often included in the formulation of the mixing length l_k.

The empirical K parameterizations are restricted by the assumption that turbulence is in the stationary balance with the mean flow. This restriction is avoided in semiempirical formulations for K_ϕ:

$$K_\phi = c_k l_k E^{1/2},\qquad(2.9)$$

where $E = 0.5(\overline{u'^2} + \overline{v'^2} + \overline{w'^2})$ is turbulent kinetic energy obtained from a prognostic equation. This type of parameterization is related to higher-order closure schemes and is discussed in the next section.

Some modelers try to overcome a fundamental weakness of the first-order closure—its underlying assumption that turbulence always acts to diffuse down the mean gradients. They introduce a countergradient correction, γ_{cg}, which applies only to turbulent heat flux in the convective ABL and allows for slightly stable stratification persisting with upward heat flux at the top of this layer (e.g., Bodin 1980; Therry and Lacarrére 1983):

$$\overline{w'\vartheta'} = -K_h\left(\frac{\partial \bar{\vartheta}}{\partial z} - \gamma_{cg}\right).\qquad(2.10)$$

The implicit K parameterizations are preferable in comparison to explicit formulas for application in mesoscale modeling. They not only contain fewer arbitrary assumptions but also allow one to parameterize turbulence in the whole computational domain including the internal boundary layer and free atmosphere above the ABL. In some models various types of K parameterizations are implemented for different turbulence regimes. McNider and Pielke (1981) suggested that the explicit O'Brien profile be used with a prognostically calculated ABL height in the convective mixing layer and that implicit-type formulations (2.6) be used for stable conditions and the internal boundary layer.

2.3. Higher-order closure

Many shortcomings of K-theory parameterizations are overcome by higher-order closure schemes. Examples of second-order models applied to ABL modeling are presented by Donaldson (1973), Mellor and Yamada (1974), and Wyngaard and Coté (1974). Second-order models use prognostic equations for second-order moments, and the third-order moments that appear in these equations must be parameterized. Generally the parameterization used for this purpose has the following form:

$$\overline{Fu_i'} = -K\frac{\partial \bar{F}}{\partial x_i},\qquad(2.11)$$

where F is a product of two turbulent fluctuations and the K coefficient may be a rather complex function of various second-order correlations. The third-order closure by André et al. (1978) is derived from the quasi-normal approximation where the fourth-order moments are expressed as combinations of the second-order moments and a damping mechanism prescribes an upper limit to the third-order moments.

Mellor and Yamada (1974) proposed a hierarchy of second-order turbulence closure models where complex model equations were systematically simplified. Their level 4 model contains full prognostic equations for all second-order moments. The level 3 model retains only prognostic equations for TKE and variances of potential temperature $\overline{\vartheta'^2}$ and humidity. In the level 2 model all prognostic equations for second moments are reduced to algebraic equations. The level 2 model and its further simplification, the level 1 model, correspond to K-theory parameterizations. Eddy diffusivities can be also derived from the level 3 model, but the turbulent heat flux is expressed in this approximation in the form (2.10) with the countergradient correction. A good compromise between accuracy and efficiency is provided by the level 2.5 model, which retains the prognostic equation for TKE only:

$$\frac{dE}{dt} = \frac{\partial}{\partial z}K_e\frac{\partial E}{\partial z} + P_s + P_b - \varepsilon,\qquad(2.12)$$

where $P_s = K_m S^2$ is shear production, $P_b = -\beta K_h \partial\bar{\vartheta}/\partial z$ is buoyancy production, $\varepsilon = E^{3/2}(c_\varepsilon l_\varepsilon)^{-1}$ is the dissipation of turbulent energy, and horizontal diffusion terms are neglected for simplicity. Eddy diffusivities for momentum, K_m; heat, K_h; and turbulent energy, K_e are expressed in the form (2.9)

$$K_m = S_m l(2E)^{1/2},\quad K_h = S_h l(2E)^{1/2},$$
$$K_e = S_e l(2E)^{1/2},\qquad(2.13)$$

where S_m, S_h, and S_e are complicated functions of wind shear and potential temperature gradient and l is the master length scale.

A necessary requirement in any turbulence model is that the resultant fields can be realized. The concept of realizability introduced by Schumann (1977) requires that the model integration ensures both that velocity and potential temperature variances remain nonnegative and that Schwartz inequalities remain satisfied. All fluxes must vanish at the same time that TKE vanishes, which means that the parameterization must exhibit a well-defined critical Richardson number. In some circumstances, use of the original level 2.5 model can lead to nonphysical results. The physical realizability of this closure scheme is discussed by Mellor and Yamada (1982), Hassid and Galperin (1983), Flatau (1985), Galperin et al. (1988), Helfand and Labraga (1988), and Andrén (1990). The simplest proposed improvement of the level 2.5 scheme consists of imposing some constraints on wind shear and the potential temperature gradient. However, this approach still does not guarantee physical realizability in the case of growing turbulence, when $E < E_r$, where E_r is the equilibrium value of turbulent kinetic energy under conditions of an exact balance between its production and dissipation:

$$P_s + P_b = \varepsilon. \qquad (2.14)$$

The fundamental assumption of the level 2.5 closure scheme—that the growth rates, advection rates, and vertical diffusion rates of the second-order turbulent moments are small and negligible—may be not valid in this case and may result in the failure of this closure. Helfand and Labraga (1988) suggested that the neglected growth rates of the second moments can be parameterized by making assumptions that the ratio of production to dissipation for each second turbulent moment is the same as it is for TKE. This means that when the level 2.5 turbulent energy E is less than E_r, the level 2.5 second moments are replaced by their level 2 counterparts, multiplied by the factor E/E_r. The level 2 scheme is based on the diagnostic equation for turbulent energy (3.14). There is no need to introduce any modifications into the original level 2.5 scheme for the case of decaying turbulence, when $E > E_r$.

The level 2.5 and 3 parameterizations, with some modifications, have been applied to many mesoscale environmental studies (e.g., Arritt 1987; Tjernstrøm 1987; Enger 1990; Uliasz 1993). Corrections proposed by Andrén (1990) for the level 2.5 model take into account the influence of the underlying surface. An important advantage of the above second-order closure is that it allows one to determine a variety of turbulent variables including variances of wind velocity components, $\overline{u'^2}$, $\overline{v'^2}$, and $\overline{w'^2}$, required as input data by air pollution dispersion models. There are many other examples of turbulence parameterizations based on the prognostic TKE equation developed using somewhat

different assumptions than the Mellor–Yamada closure technique and applied in mesoscale or regional models (e.g., Bougeault and Lacarrére 1989; Benoit et al. 1989). This parameterization may be classified as *one-and-a-half-order closure*. Applications of third-order closure models in mesoscale studies have been limited so far to idealized situations—for example, Briere (1987) studied energetics of sea-breeze circulation using a two-dimensional version of the model of André et al. (1978).

2.4. Turbulent length scale formulation

A formulation for a characteristic turbulent length scale is a common closing assumption for all implicit K-theory parameterizations and higher-order closure schemes. It is necessary to distinguish between the mixing length scale, l_k in (2.6), responsible for vertical turbulent mixing, and the dissipation length scale, l_e in (2.12), responsible for viscous dissipation (Therry and Lacarrére 1983). Several different length scales appear in the higher-order closure schemes to characterize dissipation, diffusion, and redistribution of turbulent energy, but they all are assumed to be related to a single master length scale l.

A popular approach in mesoscale modeling is to apply a simple interpolation formula proposed by Blackadar (1962):

$$\frac{1}{l} = \frac{1}{\kappa z} + \frac{1}{l_\infty}, \qquad (2.15)$$

where $\kappa = 0.4$ is the von Kármán constant. Following classical hypothesis of Prandtl, the length scale l is proportional to height $l = \kappa z$ in the surface layer where the size of the turbulent eddies is reduced by the presence of the ground surface and is limited by a constant value, l_∞, aloft. An even simpler formulation for the length scale $l = \min(\kappa z, l_\infty)$ was used in mesoscale simulations by McNider and Pielke (1981). The value of l_∞ can be parameterized in several ways (e.g., Sorbjan 1989). In the models using the prognostic equation for TKE, the expression proposed by Mellor and Yamada (1974) is frequently utilized:

$$l_\infty = \alpha \frac{\int_0^H z E^{1/2} dz}{\int_0^H E^{1/2} dz}, \qquad (2.16)$$

where H is the modeling domain height, and $\alpha = 0.1$ is a constant determined by the authors from numerical experiments. This parameter was also prescribed as a function of a local Rossby number (Hassid and Galperin 1983). The above formulation overestimated the length scale under stable conditions where the growth of the eddies is impeded by buoyancy forces. Therefore, it is necessary to introduce an upper limit for the length scale in the stable layer above the atmospheric boundary layer (André et al. 1978):

$$l = l_B \leqslant 0.75 \frac{(2E)^{1/2}}{\omega_B}, \qquad (2.17)$$

where ω_B is the Brunt–Väisälä frequency. The buoyancy length scale l_B is obtained by assuming balance between inertia represented by TKE and buoyancy forces.

Zilitinkevich and Laikhtman (1965) introduced an implicit formulation of l based on a generalized von Kármán expression:

$$l = -2\kappa\Psi\left(\frac{\partial\Psi}{\partial z}\right)^{-1}, \quad \Psi = S^2(1 - \alpha_T \text{Ri}). \quad (2.18)$$

This formulation with some modifications was used in many atmospheric boundary layer models developed in the former Soviet Union. The generalized von Kármán formula is discussed further by Łobocki (1992). Another implicit formulation of l has been suggested by Wippermann (1971) on the basis of the vertical distribution of stress.

Therry and Lacarrére (1983) derived relatively complicated expressions for l_k and l_e length scales using available simultaneous measurements of turbulent kinetic energy and dissipation, and the hypothesis $l_k/l_e \sim \overline{w'^2}/E$. Several master-scale formulations for different turbulence regimes were implemented in the Uppsala University mesoscale model (Enger 1986; Tjernstrøm 1987; Andrén 1990). In general, these length scales are designed using some extensions of the interpolation formula (2.15) and use height z, Monin–Obukhov length L, convective-layer height z_i, and the buoyancy length scale l_B (2.17) to characterize the size of turbulent eddies in the atmosphere.

An approach, interesting for mesoscale applications, was proposed by Bougeault and André (1986) and Bougeault and Lacarrére (1989). They postulated that for each level in the atmosphere, the length scales, l_k and l_e, can be related to the distances, l_{up} and l_{down}, that a parcel originating from this level and having an initial kinetic energy equal to the mean turbulent kinetic energy of the layer can travel upward or downward before being stopped by the buoyancy effect:

$$\int_z^{z+l_{up}} \beta[\bar{\theta}(z) - \bar{\theta}(z')]dz' = E(z), \quad (2.19)$$

$$\int_{z-l_{down}}^z \beta[\bar{\theta}(z') - \bar{\theta}(z)]dz' = E(z). \quad (2.20)$$

The major advantage of this formulation is allowance for global determination of the size of the eddies including effects of the distance from the ground and from stable layers. The vertical depth of an unstable layer covered by a strong inversion, z_i, is automatically selected as the length scale for turbulence while close to the surface; the height above the surface, z, is introduced as the length scale. In a layer of constant stable stratification, the length scale is proportional to the buoyancy length scale l_B. The turbulence length scale

l_k and the dissipative length l_e are obtained by the averaging of l_{up} and l_{down}. Bougeault and Lacarrére (1989) suggested $l_k = \min(l_{up}, l_{down})$ and $l_e = (l_{up}l_{down})^{1/2}$. This parameterization was originally introduced to simulate a stratocumulus-topped boundary layer (Bougeault and André 1986) and was later applied to the cases of orographically induced turbulence (Bougeault and Lacarrére 1989).

The length scale can also be defined by means of a prognostic equation. Such an equation was suggested from heuristic consideration by Bush et al. (1976). Mellor and Yamada (1982) derived a prognostic equation for the product of $q^2 = 2E$ and l in a more systematic way using the integral of the two-point correlation function. The closure assumptions involved in this derivation are complicated and less convincing than other assumptions in their closure schemes. Nevertheless, the simplified second-order closure scheme, termed q^2–l or E–l, model based on the prognostic equations for turbulent kinetic energy and length scale was used successfully in several mesoscale and regional simulations over complex terrain (e.g., Yamada et al. 1989).

An alternative approach, called E–ε closure (or K–ε in engineering applications), avoids uncertainty involved in determination of the length scale by including a prognostic equation for the dissipation rate ε in addition to the equation for turbulent energy e. The eddy diffusivities are parameterized as $K \sim E^2/\varepsilon$, and the remaining length scales can be determined hopefully more accurately from prognostic variables: $l \sim E^{3/2}/\varepsilon$. The E–ε closure has been applied in several boundary-layer studies (Deterling 1985; Andrén 1991) and in mesoscale simulations including sea-breeze circulations (Kitada 1987), effects of topography (Beljaars et al. 1987; Huang and Raman 1989), and air–sea interactions (Ly 1991). However, some modelers (e.g., Mellor and Yamada 1982) have questioned the use of the dissipation equation to determine the mixing length of the energy-containing eddies, l_k.

2.5. Nonlocal closure

Nonlocal descriptions of turbulent mixing have had a long history under many different names by many investigators (e.g., Berkowicz and Prahm 1979; Fiedler 1984; Estoque 1968; Klemp and Lilly 1978; Stull 1984) and were reviewed recently by Stull (1993). The transilient turbulence theory developed by Stull (1984) has been most extensively tested and applied in atmospheric and oceanic simulations.

In contrast to local diffusivity methods such as the K-theory closure, nonlocal closure is based on the concept that all turbulent eddies can transport fluid elements across finite distances. This idea can be explained by considering the mixing of conservative quantity ϕ in a vertical column of air between grid points i and j:

$$\bar{\phi}_i(t + \Delta t) = \sum_{j=1}^{n} c_{ij}\bar{\phi}_j(t), \qquad (2.21)$$

where the matrix of transilient coefficients c_{ij} describe mixing between all possible pairs of grid boxes. The first subscript i represents the destination grid box, and the subscript j represents the source grid box. The transilient coefficients must obey mass conservation, $\sum_{j=1}^{n} c_{ij} = 1$, for each grid box i, and tracer conservation, $\sum_{i=1}^{n} c_{ij} = 1$, for each tracer. The magnitude of c_{ij} is proportional to the amount of air undergoing mixing, while its distance from the main diagonal of the matrix is proportional to the size of the eddy. This distance is also related to the net advection speed of air parcels, because the matrix applies over a specified time step. The closure problem consists in determination of transilient coefficients. Large-eddy simulation (LES) studies of convective mixed layer and neutral boundary layer (Ebert et al. 1989) were used for these estimations, whereas approximations based on nonlocal analogies of Richardson bulk number (Stull 1993) and the turbulent kinetic energy equation (Stull and Driedonks 1987) were proposed for applications. In addition to several one-dimensional atmospheric boundary layer studies, the transilient turbulence theory was implemented in the three-dimensional Pennsylvania State University–National Center for Atmospheric Research Mesoscale Model and verified against two datasets from case studies (Raymond and Stull 1990).

2.6. Surface-layer parameterizations

Most mesoscale models have relatively low vertical resolution, which makes it difficult or impossible to directly apply first- or higher-order closures close to the ground surface. Therefore, turbulence fluxes in the surface layer are parameterized instead with the aid of the bulk aerodynamic formulation

$$\overline{(w'\phi')}_0 = -C_\alpha \bar{U}_1(\bar{\phi}_1 - \bar{\phi}_0), \qquad (2.22)$$

where C_α is a bulk transfer coefficient, $\alpha = m$ for momentum, $\alpha = h$ for heat or moisture flux, $\bar{U}$ is wind velocity, and indices 1 and 0 denote variables at the first computational level in the model and at the roughness level z_0. Assuming that the first model level is located within the atmospheric surface layer, the bulk transfer coefficients may be easily derived from the surface-layer similarity theory.

In the atmospheric surface layer with height varying from about 10 to 100 m, turbulent fluxes can be assumed to be constant:

$$\overline{u'w'} = u_*^2 \cos\mu, \quad \overline{v'w'} = u_*^2 \sin\mu,$$

$$\overline{w'\theta'} = u_*\theta_*, \quad \tan\mu = \frac{\bar{v}}{\bar{u}}, \quad (2.23)$$

and the veering of wind with height due to Coriolis effect can be neglected. The Monin–Obukhov empir-

ical flux-gradient relationships in an integrated form provide the mean wind, temperature, and other mean variables as functions of height:

$$\bar{U}(z) = \frac{u_*}{\kappa}\left[\ln\frac{z - z_0}{z_0} - \Psi_m\left(\frac{z - z_0}{L}\right)\right], \quad (2.24)$$

$$\bar{\theta}(z) - \bar{\theta}(z_0) = \frac{\theta_*}{0.74\kappa}\left[\ln\frac{z - z_0}{z_0} - \Psi_h\left(\frac{z - z_0}{L}\right)\right],$$
$$(2.25)$$

where L is the Monin–Obukhov length

$$L = -\frac{u_*^2}{\kappa\beta\theta_*}. \qquad (2.26)$$

The frequently used formulation of universal functions Ψ_m and Ψ_h was given by Businger et al. (1971) and then revised by different authors (e.g., Beljaars and Holtslag 1991). The surface-layer similarity theory may be classified as a *zero-order turbulence closure* since no prognostic equations, not even for the mean variables, are retained.

The above equation set is usually reduced to a transcendental equation that can be solved using an iterative algorithm (e.g., Berkowicz and Prahm 1982). To avoid unnecessary time consumption during the model integration, some authors introduce precomputed approximations of these relationships that provide explicit formulas for flux calculations (Louis 1979; Buyn 1990). Another possible solution is to integrate a higher-order closure equation set using surface-layer approximations (negligible advection, stationarity, vertical flux invariance). This approach is more consistent with higher-order turbulence closure schemes if it is used to represent turbulence aloft. Łobocki (1993) presented a procedure for algebraic reduction of the Mellor–Yamada level 2 model.

The surface-layer similarity expressions are usually used not only to define the bulk transfer coefficients but also to formulate lower boundary conditions in mesoscale models. The surface-layer parameters are often involved in parameterization of the subgrid fluxes in the whole ABL. In particular, all implicit K parameterizations utilize profiles of eddy diffusivities defined by the surface-layer similarity theory:

$$K_{m,h} = \frac{\kappa u_* z}{\psi_{m,h}(z/L)}, \qquad (2.27)$$

where ψ_m and ψ_h are universal functions representing nondimensional gradients of wind velocity and temperature. The surface similarity expressions should be used together with a parameterization of the viscous sublayer relating temperature and specific humidity at roughness level with their skin surface values, which is equivalent to assuming different roughness levels for wind, temperature, and humidity (Pielke 1984).

Surface-layer similarity investigations have concentrated on horizontally uniform, clear sky, flat terrain measurements. Little is known about accuracy of this theory in varying meteorological conditions and complex terrain. However, the relative success of mesoscale models using this parameterization for complex terrain simulations suggests that it still may be applicable even under complicated conditions.

Natural or man-changed land surfaces are usually heterogeneous over the resolvable scale of the mesoscale atmospheric models. Therefore, an assumption of surface homogeneity within one grid cell of the model may not represent the surface forcing accurately. A new methodology developed recently for mesoscale modeling (Avissar and Pielke 1989; Kimura 1989; Claussen 1991; Uliasz and Pielke 1992) partially overcomes this problem. Each surface grid cell of the numerical model is divided into homogeneous subregions. Under the assumption that horizontal fluxes between subgrid regions are small as compared with the vertical turbulent fluxes, patches of the same land-use type located within one grid cell are aggregated into one subgrid surface class. Then, for each of the subgrid surface classes, a micrometeorological model is applied to calculate surface turbulent fluxes of momentum, heat, and moisture. The parameters of the atmosphere—wind speed U_1, potential temperature θ_1, and specific humidity q_1—are taken from the first computational level of the atmospheric model and are assumed to be averaged over the whole grid-cell area. Then, the grid-averaged surface fluxes are obtained by the average of surface fluxes on each land-use surface weighted by its fractional area f_i:

$$-(\overline{w'\phi'})_0 = \sum_i f_i C_\alpha^i \overline{U_1}(\overline{\phi_1} - \overline{\phi_0^i}), \quad (2.28)$$

where index i denotes values for the ith land-use category. The areally averaged surface fluxes are estimated at a so-called blending height h_b, which leads to a new formulation of the bulk transfer coefficients C_α^i (Claussen 1991; Uliasz and Pielke 1992). This concept is based on the assumption that at sufficiently large heights above heterogeneous surface, subsequent surface modifications will not be recognizable in the flow individually, but overall flux and mean profiles will represent the surface conditions of a large area. The height at which the flow becomes approximately independent of horizontal position is called blending height. Below the blending height, the flow is in equilibrium with local land-use patches. The blending height is expected to vary as a function of the spatial sizes of the land surface heterogeneity and the prevailing wind speed. This approach is independent of the micrometeorological submodel used in the mesoscale model to calculate surface turbulent fluxes for different land-use patches.

Application of Monin–Obukhov theory is limited to the atmospheric surface layer, which typically represents the lowest 10% of the ABL. In some mesoscale models, the first computational level may be located above the surface layer. This problem is likely to arise under stable conditions when the surface layer may be very shallow. The bulk transfer coefficients in such a case can be determined with the aid of the *local similarity theory,* which allows one to evaluate profiles of mean variables and higher-order moments within the entire ABL (Sorbjan 1989). Local similarity functions are defined in a manner analogous to Monin–Obukhov universal functions. The appropriate scales for the stable continuous regime of the ABL are based on local (height dependent) values of momentum and heat turbulent fluxes, $u_*^2(z)$ and $\vartheta_*(z)$ (Nieuwstadt 1984; Sorbjan 1986; Sorbjan 1987).

2.7. Conclusions

A large variety of methods is currently used to deal with turbulence in mesoscale models. Recent advances in computer technology allow one to apply more sophisticated parameterizations. Although higher-order closure models are complex in their formulations, they are computationally fast enough to be useful even for operational applications. At the same time, significant progress in mesoscale modeling has been achieved by increasing vertical and horizontal resolution of numerical models. With the aid of a nested-grid technique it is possible to use parameterizations of various complexity in different parts of a mesoscale modeling domain. The approach most popular today in mesoscale applications is to use a simplified second-order closure scheme based on a prognostic equation for turbulent kinetic energy. Many mesoscale simulations have demonstrated that even the simple first-order parameterizations allow one to correctly reproduce the mean flow. However, more advanced turbulence parameterization becomes more important if the output of the meteorological model is used to run an air pollution dispersion model. Third-order closure models together with LES models that use grid-volume averaging should be considered a basic research tool due to their large computer demands. The LES research has often been performed for the idealized situation of a simple convective boundary layer over homogeneous flat terrain. However, most recent LES studies have included more complicated scenarios of horizontally inhomogeneous topography (Walko et al. 1992) or surface heating (Hadfield et al. 1992), and sloping terrain (Schumann 1990), as well as real atmospheric and terrain situations (Costigan and Cotton 1992). The last work demonstrated that LES, nested within larger grids, can be applied for case studies in complex terrain.

Chapter 3

Modeling of Surface Effects on the Mesoscale

THOMAS T. WARNER

Department of Meteorology, The Pennsylvania State University, University Park, Pennsylvania

3.1. Introduction

Mesoscale meteorological processes may originate from a variety of mechanisms; however, one of the most important is forcing by the earth's surface. Terrain height variations and differential surface fluxes of heat, momentum, and moisture force myriad phenomena on a wide range of scales—from the mesoalpha to the mesogamma (length scales of greater than 2 km and less than 2000 km). These include thermally forced circulations over differentially heated land surfaces, drylines that can be associated with spatial variations in surface moisture fluxes, mountain lee waves, mountain–valley circulations, and coastal frontogenesis and cyclogenesis.

Even though present computer systems permit the routine use of mesoscale models for operational forecasting and research applications, the simulation of surface-forced processes on the mesoscale presents special problems. That is, to correctly simulate surface fluxes of heat, momentum, and moisture, the mesoscale models must include submodels that define subsurface fluxes of heat and moisture as well as the effects of vegetation on these fluxes. In addition, these models require extensive databases about surface soil and vegetation characteristics, which can greatly complicate the process of model initialization.

The simulation of surface-forced processes also requires the use of planetary boundary layer parameterizations that are sophisticated and possess relatively high vertical resolution. The vertical distribution of the flux convergences must be accurately simulated by the turbulence parameterization in order for the correct atmospheric response to result, and the fact that the atmospheric response is often shallow means that the model's vertical resolution in the boundary layer must be high.

This chapter will 1) summarize the effect of surface-forced processes on mesoscale predictability, 2) briefly note how the feedback timescale between the surface and atmosphere determines the required sophistication of the surface-process treatment, and 3) describe two examples of how mesoscale prediction models have been used to provide a better understanding of how

surface-forced circulations interact with large-scale weather.

3.2. Surface forcing and mesoscale predictability

Anthes and Baumhefner (1984) were among the first to discuss mesoscale predictability. Other general studies address how predictability can be quantified on the mesoscale (Anthes 1984) and how it is related to uncertainty in mesoscale model initial conditions (Errico and Baumhefner 1987; Warner et al. 1984, 1989). In addition, the latter two studies illustrate how the existence of resolved local forcing will generally increase mesoscale predictability.

Local forcing has at least two effects on predictability. First, the resistance formulations that are often used to parameterize surface fluxes of heat, momentum, and moisture can cause the model solution at low levels to adjust toward a more correct value if it is not compatible with accurately specified surface conditions. Secondly, differential fluxes can cause the model to develop realistic mesoscale circulations that are not resolved by the initial data. Of course, local forcing will only reduce the error if the surface parameters are specified accurately and could increase the error if the parameters are poorly specified. The extent to which simulated local forcing influences error growth depends on whether the initial data contain information about the mesoscale locally forced component of the flow. If the forecast is initialized with information about the large-scale flow only, the mesoscale forcing will normally generate realistic mesoscale atmospheric features during the course of the integration. This effect contributes to an *increase* in the mesoscale predictive skill of the simulation during the time when the large-scale flow is responding to the local forcing. After the mesoscale features have developed, the forcing will provide a constraint on the error growth. However, when the initial state contains mesoscale detail, little increase in forecast skill will be evident, but the existence of the forcing will still limit error growth.

Figure 3.1, from Anthes and Baumhefner (1984), illustrates schematically the influence of local forcing on model predictive skill. The ordinate quantifies fore-

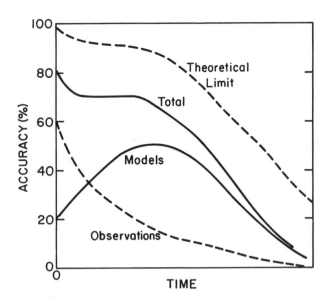

FIG. 3.1. Schematic diagram illustrating the relative contributions of observations and mesoscale model forecasts to resolving the structures of atmospheric phenomena (from Anthes and Baumhefner 1984).

cast accuracy in terms of a percentage, where 100% is a perfect forecast and 0% reflects a forecast with no skill. As shown by the dashed line, the mesoscale predictive skill based on the use of observations alone (persistence) is relatively high for short forecast periods, but it decreases rapidly. The mesoscale forecast skill is less than 100% at the zero time because the observations referred to are presumed to be too coarse to reflect the details of the mesoscale structures. On the other hand, mesoscale model forecast skill temporarily increases with time, where the most important reason for this is the response of the model atmosphere to mesoscale surface forcing that is prescribed accurately. The model skill eventually decreases with time because Anthes and Baumhefner assumed that the model errors that are related to the physical-process parameterizations, numerics, and initial conditions would eventually dominate the solution. The curve labeled "total" shows the accuracy of a forecast produced using both the data and the model prediction as guidance. The theoretical limit reflects what is attainable with mesoscale data and improved mesoscale models.

An example of the influence of surface forcing on mesoscale predictive skill can be seen in Fig. 3.2 and Table 3.1, which illustrate how surface forcing can compensate for poor initial data in simulations of the marine boundary layer structure over the Gulf Stream. These simulations employed a grid increment of 60 km and 14 computational levels. Figure 3.2 shows the 975-mb wet-bulb potential temperature field at the 12-h [panel (a)] and 20-h [panel (c)] simulation times from the control simulation of a coastal cyclogenesis event. To examine the degree to which surface forcing

could compensate for poor initial data, all the simulated variables at the 12-h time were smoothed by extracting the simulated values of the variables at every sixth grid point (i.e., 1/36 of the values were used) and then objectively analyzing the "data" back to the complete grid. The resulting smooth wet-bulb potential temperature field shown in Fig. 3.2b was then used as initial data for a 24-h simulation, the results of which are shown in Fig. 3.2d (experiment A1) for the 8-h time. It can be seen that the marine boundary layer thermal and moisture fields almost fully recovered during this relatively short period. An analysis of the dynamical processes showed that the erroneously cool and dry bias in the tongue of low-level air over the Gulf Stream that resulted from the poor resolution of the extracted information caused enhanced fluxes of heat and moisture at low levels. This surface forcing dramatically compensated for the error in the initial state. Table 3.1 illustrates the speed with which the surface forcing improved the quality of the simulation and also better quantifies the degree of the impact. The forecast times given in the table are relative to the start time of the A1 simulation at the 12-h point of the control simulation. The first quantity tabulated, the strength of the wet-bulb potential temperature gradient, represents an average over a distance of about 100 km. Note that the poor data resolution caused the gradient to be too weak [$3.2°C (100 \text{ km})^{-1}$ in A1 versus $6.6°C (100 \text{ km})^{-1}$ in the control]. After the first 4–8 h of the A1 simulation, the baroclinic zones have very similar strengths. The second parameter shown is the maximum latitude of the 16°C isotherm, which reflects the extent of the northward penetration of warm moist air along the coastline associated with the cyclone. Again, the error is virtually eliminated during the first 4 h of the A1 simulation. As noted earlier, the degree to which the surface forcing improves the predictability depends on the accuracy with which the surface conditions are specified; in this case the relevant surface condition is the sea surface temperature.

3.3. Feedback between surface forcing and mesoscale processes

In relatively simple numerical models of mesoscale processes, many of the surface characteristics are held constant and are not allowed to respond to atmospheric processes. For example, even though an energy budget equation may be used to predict the spatial and temporal variation of the surface temperature, parameters such as soil moisture may be held constant with time. However, in more complete treatments of atmosphere–surface interaction, the surface conditions must be allowed to respond to the atmospheric processes so that their continuing influence on the atmosphere can be better simulated. For example, simulated precipitation must be allowed to modify the soil moisture and the state of the vegetation, and the insolation must be based on the existence of simulated cloud and fog.

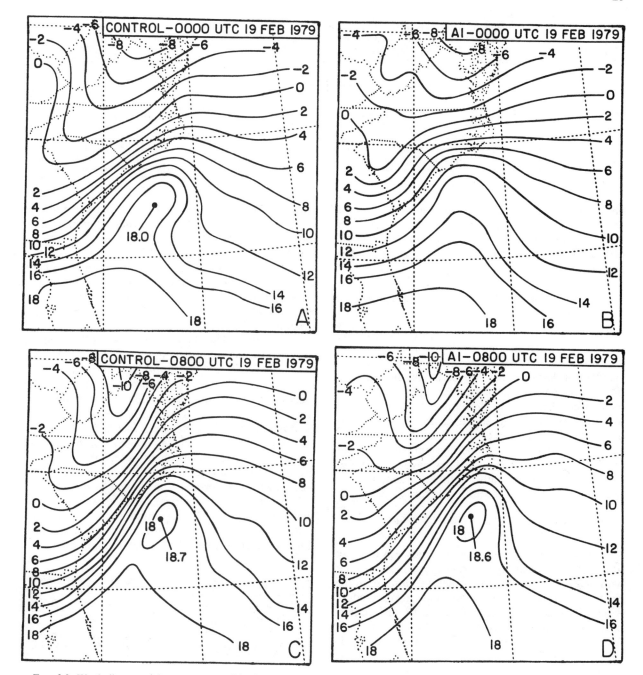

FIG. 3.2. Wet-bulb potential temperatures (°C) from a control simulation and experiment A1, where the A1 experiment involved smoothing of all variables and reinitialization of the model at the 12-h time of the control simulation: (a) control simulation at 12 h, (b) smoothed field from the 12-h control simulation, (c) control simulation at 20 h, (d) simulation 8 h after the restart of the model from the smooth field shown in panel (b). The isotherm interval is 2°C.

Naturally, the importance of the surface–atmosphere feedback to the forecast accuracy depends on the timescale of the feedback processes relative to the length of the forecast. That is, the veracity of extremely short range forecasts of a few hours will be dominated by the accuracy of the initial conditions and will not depend greatly on longer-timescale surface–atmosphere feedback processes. An example of a short-timescale

feedback process is the wetting of foliage during a precipitation event and the rapid subsequent evaporation from the foliage to the atmosphere. In contrast, a longer-timescale feedback process would involve the percolation of precipitation into the plant root zone, with the resulting moistening of the boundary layer by evapotranspiration. Over coastal zones, a short-time-scale feedback would be the influence of wind speed

TABLE 3.1. Error evolution of the wet-bulb potential temperature pattern along the Carolina coast. The forecast time is relative to the time when the simulation was started from the smoothed conditions. The control simulation was initialized 12 h prior to this time. Experiment A1 involved the restart of the model after all variables were smoothed as described in the text.

Feature	Forecast time (h)						
	0	4	8	12	16	20	24
Maximum θ_w gradient—control [°C (100 km)$^{-1}$]	6.6	8.5	8.0	9.0	7.2	5.1	4.9
Maximum θ_w gradient—Al [°C (100 km)$^{-1}$]	3.2	7.3	8.0	8.2	7.2	4.3	4.5
Maximum latitude of the 16°C isotherm—control (°N)	33.1	33.7	34.2	34.6	34.3	33.9	34.0
Maximum latitude of the 16°C isotherm—Al (°N)	30.9	33.6	34.1	34.6	34.1	34.0	34.0

on ocean wave characteristics, with the resulting influence on roughness length and evaporation rate. A somewhat longer timescale process would involve the feedback between the atmospheric and ocean boundary layer circulations.

3.4. Examples of mesoscale modeling of surface forcing

There have been a very large number of modeling studies performed that deal with the response of the atmosphere to surface forcing. Some have been motivated by a desire to develop better prediction techniques for operational forecasting applications, some have been designed to develop mesoclimatologies, while others have been aimed at improving our understanding of the physical processes themselves. Because these studies are so numerous, only two examples will be briefly described here. They have been chosen because they represent rather complex interactions between the surface and the atmosphere.

a. Mesoscale circulations forced by boundaries between irrigated and nonirrigated surfaces

This observational and modeling study was designed to investigate thermally forced circulations associated with irrigated crop areas in northeast Colorado. In this geographic area, irrigated crop lands are embedded within dry land, with satellite-derived IR surface temperatures being about 10 K lower over the irrigated surface. Aircraft and surface measurements also indicated a significantly lower air temperature and greater moisture content at low levels over the moist surfaces. The model simulations complemented the measurements and were used to study the generation of the thermally forced flows and their impact on the large-scale flow field.

The numerical model formulation is described in Pielke (1974), Mahrer and Pielke (1977a), and McNider and Pielke (1981). Even though the model normally employs a surface energy equation for the calculation of ground temperature, in this case the computed surface temperatures were replaced with satellite-based estimates of the surface temperature. This version of the model employed a grid increment of 6.9 km and 20 computational layers. The simula-

tions were initialized at 0600 MDT with three different large-scale wind regimes, a zero mean synoptic flow, and northeasterly and southerly synoptic flows of 2.5 m s^{-1}.

Figure 3.3 shows the computational domain of the model simulations and indicates the topography as well as the distribution of the irrigated areas (shaded). The results of the simulation with no large-scale mean flow are shown in Fig. 3.4, which illustrates the surface sensible heat flux (Fig. 3.4a), the surface (5-m height) horizontal wind velocity (Fig. 3.4b), and a west–east vertical cross section of the horizontal wind velocity across the middle of the computational domain. The effect of the irrigation on the sensible heat fluxes is obvious; over the irrigated areas in the eastern part of the domain and along the South Platte River (see Fig. 3.3), the sensible heat fluxes range from 200 to 300 W m^{-2}, in contrast to fluxes in excess of 400 W m^{-2} elsewhere. The surface-layer wind field in Fig. 3.4b shows significant perturbations forced by the domain-scale gradient in surface heat fluxes as well as by the finer-scale contrasts. For example, note the existence of the lower branch of a thermally direct circulation along the eastern one-third of the domain, especially

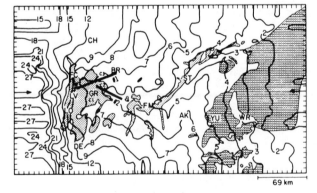

FIG. 3.3. Model computational domain showing the topography and the irrigated areas (stippled). The irrigated area along the South Platte River extends from the western third of the domain eastward with an irregular pattern until it intersects the large area in the eastern third of the domain, which is semi-irrigated farmland. The terrain elevation is given in hundreds of meters above the lowest elevation of the domain (820 m MSL). The thick lines represent aircraft flight paths (from Segal et al. 1989).

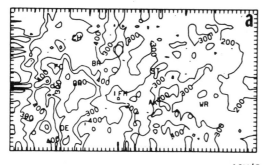

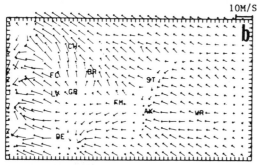

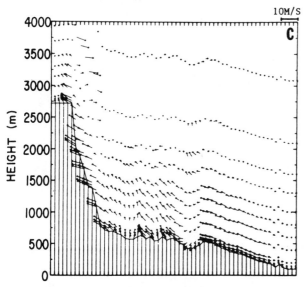

FIG. 3.4. Numerical model–simulated fields for 1400 MDT with no mean synoptic flow: (a) surface sensible heat fluxes (W m^{-2}); (b) surface horizontal wind velocity (at 5 m above ground); and (c) east–west vertical cross section through the middle of the domain showing the horizontal wind velocity (upward velocity component indicates southerly flow) (from Segal et al. 1989).

between locations AK and WR, which are on opposite sides of the irrigation-area border. In addition, there is a significant north–south-oriented diffluent flow at low levels away from the cooler irrigated area along the South Platte River. Note that the westward-oriented gradient in the elevation of the heated surface contributes to the development of the mean eastward flow over the domain at low levels, which reinforces the forcing associated with the greater amount of irrigation in the eastern portion of the domain.

The cross section of the wind field shown in Fig. 3.4c, representing a west–east transect of the domain, shows that the flow field perturbations forced by the surface-moisture contrasts and topography extend through a significant depth in the boundary layer. For example, the relatively calm conditions seen between AK and ST in Fig. 3.4b at the western edge of the thermal circulation, and the easterlies over the eastern third of the domain, extend through a deep layer. Also, many small-scale variations in the west–east wind component are seen where the cross section intersects the varied irrigation pattern to the east of GR along the South Platte River.

This modeling study not only shows the characteristics of an interesting surface-forced mesoscale circulation, but it also illustrates how a mesoscale model can be used to determine the impact of existing or proposed land-use changes on the mesoclimate.

b. Precipitation production along the edge of an elevated mixed layer triggered by surface forcing

An 18-h numerical simulation of the weather associated with the severe storm outbreak in the region of the Texas–Oklahoma panhandles during the AVE-SESAME IV (Atmospheric Variability Experiment–Severe Environment Storms and Mesoscale Experiment) study period (9–10 May 1979) was performed (Lakhtakia and Warner 1987). This simulation and the related sensitivity tests provided the four-dimensional datasets that were used to reach a better understanding of the processes that were involved, in this case, in the development of severe convection along the edge of an elevated mixed layer (EML).

Elevated mixed layers are frequently observed in the Great Plains of the United States and are often responsible for establishing the large-scale environment for severe convective storms. Carlson and Ludlam (1968) and Carlson et al. (1983) describe the conceptual model of the EML (illustrated in Fig. 3.5). This mixed layer has been advected over the lower terrain of the southern Great Plains from the elevated heat source of the Mexican Plateau. Its dynamical importance relates to the fact that its lower boundary forms a layer of very strong stability that acts as a lid or upper boundary to convection in the moist, potentially cooler, tropical air from the Gulf of Mexico that flows northward at low levels. Therefore, the EML causes the buoyant energy to build up in the boundary layer. As a result, severe convection can occur along the northwestern boundary of the EML (or lid edge) if the low-level air runs out from beneath it in a process called underrunning.

The sensitivity studies were performed to isolate the contribution of differential surface forcing to the development of underrunning, the intensification of the underrunning, and the resulting heavy rainfall. It was hypothesized that, due to differential surface heating

FIG. 3.5. Schematic description of the severe weather environment over the southern Great Plains of the United States. Above the topography, an EML is shown on the right and a stationary front on the left. The location of the convective outbreak is represented by a cumulonimbus between the EML and the front (from Lakhtakia and Warner 1987).

near the EML edge, a slight underrunning wind component appeared that caused weak moist convection that induced further low-level convergence, which in turn forced a stronger underrunning component that further enhanced the convection. To test this hypothesis, two simulations were performed—one with a heterogeneous representation of surface characteristics, and one with horizontally homogeneous values for all the parameters (except for terrain elevation). That is, in the latter case average parameter values were used, which essentially produced horizontally uniform surface fluxes of heat and moisture.

The numerical model used for this study was a version of the Pennsylvania State University–National Center for Atmospheric Research Mesoscale Model (Anthes and Warner 1978). This three-dimensional, hydrostatic model employed a high-resolution planetary boundary layer parameterization, a surface energy budget equation for ground temperature, and a convective parameterization of the Anthes–Kuo type (Anthes 1977) to simulate convective precipitation. The grid increment was 30 km. The predictive equation for ground temperature employed horizontally variable values of surface albedo, soil moisture availability, thermal inertia, surface emissivity, and roughness length. Soil moisture availability was specified based on climatology and antecedent precipitation. Cloud fraction was diagnosed using an empirical relationship between cloud cover and relative humidity, and the cloud cover information was then used to modify the solar and terrestrial radiative fluxes at the surface that were used in the surface energy budget equation.

During the first few hours of the simulation, the EML's effect of inhibiting deep mixing promoted an increase in relative humidity under it, which resulted

in the diagnosis of low-level clouds by the model to the east of the lid edge. During the same period, the area to the west of the EML and in the vicinity of the lid edge was simulated to have mostly clear skies. This simulated cloud cover gradient is consistent with the observations. Thus, after sunrise there was an increasing difference in the insolation between the area under the EML and the area to the west of it. Another factor that reinforced this differential surface-heating pattern was the lower soil moisture availability that existed in western Texas compared to eastern Texas. Therefore, the higher ground temperatures in western Texas led to the transfer of more sensible heat to the atmosphere. Figure 3.6 shows this effect in terms of the model-simulated surface temperature for the lowest model level and the surface sensible heat flux at 2100 UTC. The distinction is obvious between the area with clear skies and the area partially covered with low-level clouds.

To illustrate the spatial and temporal relationship among the low-level wind direction, the lid-edge position and orientation, and the precipitation rate, Figs. 3.7 and 3.8 show the variation with time of the zonal distribution of the simulated wind vectors on the 0.965 sigma surface and the 3-h simulated precipitation for the control simulation and the horizontally uniform surface flux experiment, respectively. It is clear from Fig. 3.7 that some underrunning was present at 2100 UTC in the control simulation and that the convective outbreak started at near this time. However, Fig. 3.8 shows that, with horizontally uniform fluxes, the underrunning component in the low-level winds was extremely weak or nonexistent and that the 3-h precipitation rates were much lower than the ones predicted in the control simulation. The spatial and temporal

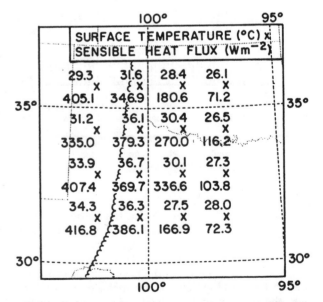

FIG. 3.6. Model-predicted surface temperature and sensible heat flux in Texas and Oklahoma at 2100 UTC. The scalloped border represents the lid edge (from Lakhtakia and Warner 1987).

distribution of the observed rainfall and that predicted by the control simulation are very similar (not shown).

This is an illustration of a scale-interaction process in which differential surface forcing on the mesoscale combined with a particular synoptic-scale stability pattern to trigger heavy convective rainfall. The differential surface fluxes of heat and moisture resulted primarily from differential cloud shading that was caused by different mixing depths underneath and to the west of the EML.

3.5. Summary

It has been shown that mesoscale predictability can be significantly enhanced through the existence of surface-forced circulations, provided that the surface fluxes

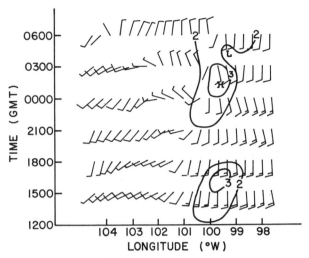

FIG. 3.8. As in Fig. 3.7 but for the simulation without differential surface heat and moisture fluxes. The lid edge is not shown, but its position is similar to that in Fig. 3.7 (from Lakhtakia and Warner 1987).

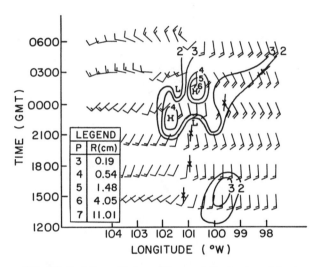

FIG. 3.7. Variation with time of the zonal distribution of the simulated wind vectors on the 0.965 sigma surface, for the total period of the control simulation (with spatially variable surface fluxes), and the precipitation accumulated over 3-h periods. Precipitation isopleths are of $P = \ln(R + 0.01) + 4.6$, where R (cm) is rainfall. Crosses indicate the position of the lid edge, and the line through each cross shows the orientation of the lid edge (from Lakhtakia and Warner 1987).

are adequately simulated. This requires the use of often complex databases that define surface properties such as soil and vegetation characteristics, and the use of models that simulate the transport of water and thermal energy at and below the earth's surface. The required sophistication of these surface-process submodels depends on the timescale of the interaction between the surface forcing and the atmosphere and its relationship to the duration of the model forecast. Two examples were used to illustrate the complex signature that surface forcing can have on the large scale, and how surface-forced circulations can serve as a trigger to release large amounts of convective energy.

The successful use of mesoscale models for operational forecasting of phenomena ranging from coastal cyclogenesis to convective storms will depend to a significant degree on our ability to simulate these surface-forced processes and the boundary layer turbulence that provides the dynamical connection between the surface and the atmosphere.

Chapter 4

Representing Moisture Processes in Mesoscale Numerical Models

JERRY M. STRAKA

School of Meteorology, University of Oklahoma, Norman, Oklahoma

4.1. Introduction

Moisture variables and associated physics are necessary in mesoscale models to properly predict virtual temperature, moisture distributions, cloud formation, precipitation formation and fallout, latent heating associated with phase changes of water, and surface fluxes of heat and moisture. Without properly simulating, for example, the diabatic heating associated with phase changes of water substance and surface fluxes of water vapor, we would not be able to accurately predict the formation and intensification of many mesoscale circulations, as well as both smaller- and larger-scale circulations. Moreover, the inclusion of moisture in mesoscale numerical models facilitates a more accurate foundation for describing the radiation balance of the atmosphere and permits cloud and precipitation forecasts.

In section 4.2, we briefly discuss some of the issues of including surface moisture physics. Further discussion on this topic is presented in chapter 3. In section 4.3, we discuss the inclusion of stratiform and convective precipitation physics in mesoscale numerical models. In section 4.4, we briefly review some of the initialization, assimilation, and retrieval techniques for improving the prediction of moisture, clouds, and precipitation, and the latent heating associated with phase changes of water substance in mesoscale models. Finally, we present a discussion and conclusions in section 4.5.

4.2. Surface moisture

A review of the literature concerning the influence of surface moisture in mesoscale numerical models reveals a number of repeated contentions. For example, statements are made that suggest that the influence of soil moisture is not studied as much as other aspects of mesoscale modeling; soil moisture and surface albedo are two of the most important variables that can control feedbacks of surface physics to the atmosphere; more than just a single representative value of soil moisture is needed for successful mesoscale prediction; often soil moisture can

influence a simulation as much as dynamical processes; and soil moisture can vary rapidly in time and feedback into the simulated (and therefore the real) atmosphere in only a few hours.

The sensitivity of boundary layer simulations to surface moisture availability, as compared to surface roughness, surface albedo, and slab thermal heat capacity, is demonstrated in Fig. 4.1 (reproduced from Zhang and Anthes 1982). The sounding used in the boundary layer model was from Marfa, Texas, at 0400 LST 10 April 1979 (Fig. 4.2). The model integrations were made assuming a steady geostrophic wind. For the range of parameters chosen, the effect of surface moisture availability is the most significant. The boundary layer is shallower, and the atmosphere is cooler and more moist for simulations using larger values of surface moisture availability. While the importance of the parameters, relative to each other, is not explicitly mentioned in the Zhang and Anthes paper, each parameter is varied over a range of realistic values.

To accurately incorporate surface physics into a mesoscale numerical model, it is necessary to use an appropriate parameterization of subgrid-scale air–sea and air–land moisture and heat fluxes. In many instances, it is essential that a comprehensive surface energy budget be used. Most of these include detailed surface and soil models (including latent and sensible heat flux, surface type and use, soil porosity, heat capacity, and moisture content) to accurately simulate the shortwave and longwave radiation processes at and near the surface. Some of the more sophisticated surface physics packages are coupled with complex vegetation and biosphere models that include plant type and associated water physics (e.g., Segal et al. 1988; Avissar and Pielke 1989; Pielke and Avissar 1990). To use such complex models, Pielke et al. (1991) have suggested that a two-dimensional spectral characterization of surface quantities that affect moisture availability—including surface albedo, soil moisture, surface roughness, soil type, and vegetation index—is needed. Care must be taken, however, so that the spectral characterization is consistent with model grid resolution.

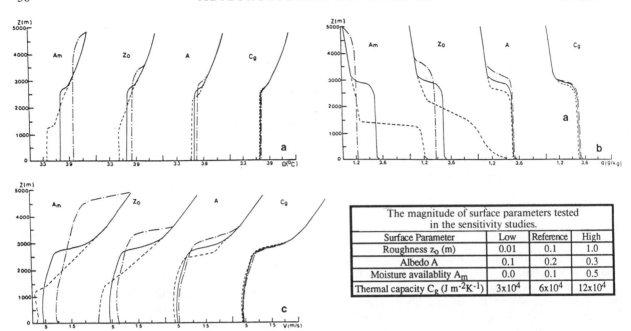

FIG. 4.1. Vertical profiles of (a) potential temperature θ, (b) specific humidity q, and (c) wind speed V at 12 h for various values of moisture availability A_m, roughness parameter z_o, albedo A, and thermal capacity C_g [as described in Zhang and Anthes (1982)]. Reference values of parameters are denoted by solid lines, high values by dashed lines, and low values by dash–dot lines as described in the figure (from Zhang and Anthes 1982).

4.3. Precipitation physics

In this section, we review the inclusion of stratiform and convective precipitation physics in mesoscale numerical models. Much of the material presented on convective precipitation physics is based on previous reviews including those by Ooyama (1982), Frank (1983; referred to as F83), Pielke (1984; referred to as P84), Arakawa and Chen (1987), Tiedtke (1988), Cotton and Anthes (1989; referred to as CA89), and Molinari and Dudek (1992; referred to as MD92), all of which are highly recommended.

Because of the horizontal grid resolutions typically used in mesoscale numerical models, the specification of precipitation formation in both stable conditions (e.g., equivalent potential temperature θ_e increases with height: $\partial\theta_e/\partial z > 0$) and unstable (or convective) conditions (e.g., equivalent potential temperature decreases with height: $\partial\theta_e/\partial z < 0$) can be a difficult problem. Following P84, stable precipitation requires sustained forced ascent of humid air to produce saturation; clouds; and, subsequently, precipitation. For unstable precipitation, lifting is only required to saturate air parcels. By some definitions, lifting is also needed for parcels to reach their level of free convection. Once a parcel is saturated and above its level of free convection, it will continue to rise on its own and produce condensate without sustained large-scale updrafts. The net effect of convection is to heat and dry the atmosphere (as well as transport momentum). The resulting diabatic heating from the formation of stable and unstable

precipitation can be essential in the development of many mesoscale weather systems. Moreover, precipitation production and mesoscale development can be enhanced through cooperative feedback mechanisms.

a. Stratiform precipitation

Stratiform precipitation is traditionally computed explicitly by determining the grid-scale relative humidity, checking to see if the relative humidity exceeds a threshold value (typically, relative humidity equal to 90%–100%), and forcing the excess moisture to condense while simultaneously warming the atmosphere so the threshold criterion is satisfied. In the most simple physical models, the condensate (i.e., the precipitation) is immediately accumulated at the ground. In slightly more complex physical models, the relative humidity is checked sequentially in the layers below, which are moistened and cooled simultaneously to the relative humidity threshold if the layers are unsaturated. Precipitation is accumulated on the ground if any remains after saturating the layers between where the condensation was occurring and the ground. This type of scheme could be described as explicit condensation physics with highly parameterized precipitation physics (i.e., zero-order precipitation physics). Examples of models using this type of treatment include nearly all of those in the past 30 years that have been operational at the National Meteorological Center (NMC) [Limited-Area Fine-Mesh Model (LFM), Gerrity 1977; Nested Grid Model (NGM), Phillips 1979, etc.] and

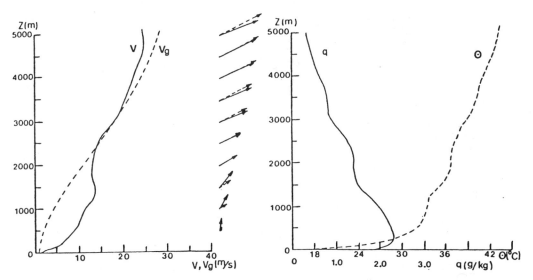

FIG. 4.2. Initial vertical profile of potential temperature θ, specific humidity q, wind speed V, and geostrophic wind speed V_g for Marfa, Texas, 1200 UTC 10 April 1979 (from Zhang and Anthes 1982).

the U.K. Meteorological Office (Gadd 1978), as well as many research mesoscale models including the early Pennsylvania State University–National Center for Atmospheric Research Mesoscale Model (Anthes and Warner 1978).

The condensation–precipitation scheme described above can be modified by including the integration of a prognostic cloud content variable (Sundqvist 1978; Golding and Ballard 1989; Zhang et al. 1988; Zhao 1993). This variable acts as a condensate storage term and can improve precipitation forecasts and radiation balance calculations. However, precipitation is still assumed to form and fall out to the ground in one time step, precluding the need for carrying prognostic equations for precipitation content. Improvements of more than 10% in precipitation forecast threat scores have been made by Zhao (1993) using this type of precipitation physics in NMC's eta model. Golding and Ballard (1989), and others, have also made similar advances.

Finally, some models use bulk explicit precipitation physics, which incorporate prognostic continuity equations for water vapor, cloud content (often using separate equations for cloud water and cloud ice), and precipitation content (often including separate prognostic equations for rain, snow, and snow aggregates, and graupel and hail). Complex interactions (mass transfers from one hydrometeor species to another by conversions, vapor diffusion, and accretion) are allowed to occur between each of the hydrometeor species. In addition, precipitation fallout is not assumed to occur simultaneously; it is accomplished by computing a tendency due to gravitational settling. Bulk explicit schemes are usually reserved for use in only the finest-scale models—such as the Colorado State University Regional Area Modeling System [CSU

RAMS, Cotton et al. (1982)] and other similar cloud models (e.g., Klemp and Wilhelmson 1978; Clark 1979; Lin et al. 1983; Straka and Anderson 1993; Johnson et al. 1993). However, Zhang et al. (1988) and others have also used this approach for higher-resolution simulations with mesoscale models.

b. Convective precipitation

There are a number of choices concerning complexity and reliability that need to be addressed for representing convective precipitation processes. In general terms, these choices involve 1) implicit representations of precipitation processes by using parameterizations where the bulk or average grid volume effects of precipitation formation are computed based on grid-scale observables, or 2) explicit (nonparameterized) representations of precipitation processes where the gross size of precipitation-producing clouds is assumed to be resolved by the model grid (e.g., P84; MD92).

With the implicit approach, it is assumed that the properties of convective clouds are different from the properties of the grid scale. Thus, the cumulus-scale physics need to be represented or parameterized in terms of the resolvable scale. Paraphrasing GARP (Global Atmospheric Research Program) Publication Series No. 8 and CA89, it is suggested that a successful parameterization requires (i) identification of the process to be parameterized, (ii) determination of the importance of the process to the resolvable scales of motion, (iii) intensive study of individual cases in order to establish the fact that the relevant physics and dynamics are adequately understood, and (iv) formulation of quantification rules for determining the grid-scale averages of the transports of mass and momentum, heat and moisture, and verification of these rules by comparison with direct observations. Furthermore, the parameterization closure should probably be unique for a given combination of grid-scale observables.

With the explicit approach, the smallest resolvable cloud is the size of a grid element; clouds are resolved by the model's grid. Prognostic continuity equations are used to explicitly represent cloud and precipitation hydrometeors, with source and sink terms accounted for by microphysical parameterizations (i.e., the parameterization is of processes on scales much smaller than the grid scale). For excellent reviews and references on explicit schemes the reader is referred to CA89 and references in section 4.2a.

A question posed to mesoscale models is whether convection should be represented explicitly or implicitly. Frank (1983) suggests examining the dynamic Rossby radius of deformation R', which, for a stable atmosphere, is defined as

$$R' = \frac{NH}{(\zeta + f)^{1/2}(2VR^{-1} + f)^{1/2}},$$

where N is the Brunt–Väisälä frequency, ζ is relative vorticity, f is the Coriolis parameter, V is horizontal velocity, R is the radius of curvature of the flow, and H is scale height. The interpretation of H could be the height of the troposphere or the characteristic depth of a convective system. The parameter R' is interpreted as the "relative resistance of vertical displacements resulting from static stability and horizontal displacements resulting from inertial stability" (F83). The value R' can similarly be described as the scale at which rotational aspects or inertial stability of a dynamical system becomes important (CA89). For $L < R'$, thermal instability dominates over rotational stability, and energy is dispersed by gravity waves in both the vertical and horizontal. If $L > R'$, the flow is nearly balanced (in the horizontal). Moreover, vertical motions are controlled by the primary circulations.

A schematic diagram depicting the relationship between R', horizontal scale length, and modes of atmospheric circulations is presented by F83 and Ooyama (1982) and is reproduced in Fig. 4.3. Frank (1983) is careful to state that physical scales and dynamical scales are different, with small dynamical scales referring to cases where $L < R'$, and large dynamical scales referring to cases where $L > R'$. Frank (1983) also uses the physical scale definitions of Orlanski (1975) in which wavelengths of 20–2000 km are defined as mesoscale, and wavelengths of 2000–10 000 km are defined as synoptic scale. In region I of the schematic presented in Fig. 4.3, phenomena in the atmosphere are considered "probabilistic" for any model with a grid resolution smaller than the horizontal scale of an individual cloud. Region II describes highly unbalanced flows where the divergent component of the

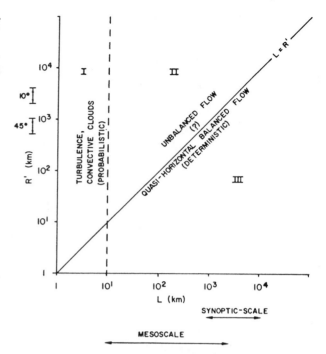

FIG. 4.3. Schematic of the relationship between horizontal scale L and Rossby radius of deformation R' and modes of circulation. Values of R' at 10° and 45° latitude are indicated (from Ooyama 1982; Frank 1983).

wind is not considered a secondary circulation. As pointed out by F83, the dynamics in this region are not well understood for the case when there are large internal energy sources. In region III, atmospheric phenomena are nearly balanced. Large-scale systems evolve slowly, and secondary circulations are controlled by the primary large-scale circulations (F83).

Frank (1983) concludes that the type of convective precipitation physics in a model must depend on the nature of the modeled circulations. The timescale of the circulations parameterized must also be taken into account. For flows in region III of Fig. 4.3, there is a well-defined scale separation. It is unlikely that convective-scale or mesoscale systems can force significant large-scale divergence, which can feed back onto the convective or mesoscale itself; rather, in this regime, it is primarily the rotational wind that adjusts to latent heating (Haltiner and Williams 1980; Shapiro and Willoughby 1982). Thus, in region III, convection appears to be modulated by large-scale thermodynamic instability and low-level forcing. Also, the large scale is insensitive to the timing of convective latent heating, which simplifies the cumulus parameterization problem (F83). In region II, the cumulus parameterization problem becomes increasingly more complex as the scale separation between convection and the large scale becomes "blurred." Cumulus parameterizations must be able to precisely determine the timing and rates of convective heating, moistening, and momentum fluxes in conditionally unstable atmospheres (F83). Toward region I, most cumulus parameterization schemes begin to break down as convection begins to be explicitly resolved by grid-scale motions.

More recently, MD92 propose that schemes for including convective effects in mesoscale models can be categorized by the following three approaches: 1) the traditional approach utilizes cumulus parameterizations at convectively unstable grid points and explicit condensation at convectively stable grid points; 2) the fully explicit approach uses explicit methods regardless of the convective stability; and 3) the hybrid approach parameterizes convective-scale updrafts and downdrafts but detrains precipitation particles to the grid scale. The horizontal grid resolutions that are most appropriate for using these approaches are summarized in Fig. 4.4 (reproduced from MD92). The following discussion on this topic is based to a large extent on MD92.

1) TRADITIONAL APPROACH

For horizontal grid scales larger than 50–60 km, MD92 suggest using cumulus parameterizations at convectively unstable grid points and explicit condensation at convectively stable grid points. Justifications for using traditional types of cumulus parameterizations include that 1) significant precipitation and tropospheric heating are produced by convection, even when the grid scale is unsaturated, and 2) large vertical eddy fluxes of heat, moisture, and momentum occur on the unresolved convective scale. Without a cumulus parameterization, model performance could be compromised when conditional instability is contained in the initial conditions or predicted because the effects associated with deep convection might not be correctly represented by the grid-resolved physics (MD92).

The assumptions imposed in developing a "traditional approach" cumulus parameterization scheme are that 1) there is a gap in the energy spectrum between the resolved scale and the parameterized scale (MD92; F83; Lilly 1960); 2) individual convective elements complete their life cycles before grid-resolved signals (MD92; F83); and 3) convection occurs in an area that is much smaller than the area allowed by the grid scale (MD92; F83).

2) EXPLICIT APPROACH

The use of explicit cumulus physics representations becomes necessary for horizontal grid resolutions less than 3 km. At this scale, large deep convective clouds are often resolvable (e.g., Lilly 1990). Arakawa and Chen (1987) suggest that a horizontal resolution as small as 100 m is needed to explicitly represent convection accurately; however, turbulence and microphysics would still have to be parameterized at this resolution.

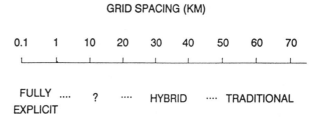

FIG. 4.4. Molinari and Dudek's (1992) proposed form for cumulus parameterization in regional mesoscale models as a function of grid spacing. The scale is logarithmic below 10 km and linear above. The question mark indicates the lack of an obvious solution, and dots represent transition regions between choices. It is assumed the model covers a wide enough area that the approach must simulate convective effects over a wide range of thermodynamic and inertial stability regimes (from Molinari and Dudek 1992).

Success with the explicit approach at grid resolutions greater than 3 km is possible for cases of modest large-scale instability, very strong forcing, and dynamically balanced systems (MD92); these conditions are typically found with tropical cyclones (e.g., Rosenthal 1978, 1979; Jones 1986; Rotunno and Emanuel 1987; MD92). Failures using the explicit approach at these resolutions can be attributed to a lag in the spinup of moisture because of lack of vertical convective fluxes (Tripoli et al. 1986; Tripoli 1986; Sardie and Warner 1985; Nordeng 1987; Dudek 1988; Zhang et al. 1988; Cram et al. 1992a,b; MD92). Once convection develops, it tends to occur over too large an area, producing excessive precipitation and unrealistic low-level heating (MD92).

A possible justification for using explicit physics might be based on whether the vertical heat and moisture fluxes on the grid-resolvable scale dominate those on the subgrid (convective) scale (P84), and whether there is a very strong coupling between the cloud scale and mesoscale, as in a strongly forced, dynamically balanced system.

3) HYBRID APPROACH

The most troublesome scales for parameterizing convective processes are those between 3 and 50 km. At these scales, the mesoscale response to convection is often resolvable [i.e., the lower- and upper-level outflows of the mesoscale convective system (Fritsch and Chappell 1980a,b) and substantial portions of the precipitation are stratiform, which can force widespread lower-tropospheric downdrafts (Zipser 1977; Houze 1989)]. Moreover, precipitation and cloud particles that are detrained from convective elements into the stratiform region are often responsible for the mesoscale organization.

To accurately simulate the evolution of the mesoscale convective system, MD92 point out that "The mesoscale modeler must simulate (i) the vertical distributions of subgrid-scale heat, moisture, and hydrometeor sources due to subgrid-scale convection; (ii) their time variation over the life cycle of the system, as mesoscale downdrafts increase in areal coverage; and (iii) the explicit updraft–downdraft couplet, saturated aloft and unsaturated below, which is left behind." Thus, more detail than the traditional cumulus parameterization approach can possibly provide is required. An alternative approach for these scales is the hybrid scheme.

The first assumption of the hybrid scheme is that the effects of implicit (parameterized) convective clouds are defined on the basis of the resolvable-scale quantities, but that the properties of convective clouds are different than the properties of the grid scale. This means a closure condition must be specified, such as large-scale conditional instability and low-level convergence. The cumulus parameterization of the hybrid approach is needed to compute the net convective-scale heating and condensate. The second assumption is that a fraction of the implicit convective condensate is "detrained" from the unresolved cumulus scale to grid-scale prognostic cloud and precipitation equations (MD82). Detrained particles then can grow in the "anvil" regions by vapor diffusion, with the addition of latent heat, thereby producing weak "anvil updrafts." As the precipitation particles fall, they can melt (or evaporate) over widespread regions if the air is warmer than the freezing temperature (or the air is unsaturated). The loss of latent heat due to melting and evaporation, along with precipitation loading, can then result in large areas of weak downdrafts.

An example of a hybrid-type scheme is that developed by Frank and Cohen (1987). Using this method, Cohen and Frank (1987) were able to simulate many features of the organization of tropical systems using a grid resolution of 25 km. Yamasaki (1987) also developed a hybrid scheme, which he used to successfully simulate a tropical cyclone. In another study, Dudhia (1989) used a two-dimensional model with 10-km resolution and the Frank and Cohen approach modified to include the ice phase to simulate convective and mesoscale features of diurnal convective events over the South China Sea. Sensitivities of the hybrid approach have been found for choices of parameters in the cumulus parameterization scheme (Dudhia 1989) and both the length and dynamical scale of the weather system modeled (MD92). A problem with the hybrid approach is that double counting of the same physical process can occur (i.e., simultaneously parameterizing and explicitly resolving), especially at higher resolutions. Another problem is that the amount of convective precipitation detrained into the environment must be specified somewhat arbitrarily (Kreitzberg and Perkey 1976; Frank and Cohen 1987; Dudhia 1989; MD92).

Nevertheless, the hybrid approach has significant advantages over both the traditional and explicit approaches in that the moisture spinup problem is alle-

viated and it contains the physics necessary to produce organized mesoscale circulations (MD92). Moreover, the hybrid approach seems to be the most superior of all approaches in separating slow and fast processes into explicit and parameterized parts for horizontal grid resolutions of 10–50 km (MD92).

c. Slantwise convection

Another concern to the mesoscale modeler is the influence of slantwise convection. It is suggested by MD92 that there are three options for simulating the influence of slantwise convection in a mesoscale model: "(i) parameterize slantwise convection separately from both upright convection and explicit grid-scale condensation; (ii) use a fully explicit approach with no parameterization of either upright or slantwise convection; and (iii) use a traditional or hybrid approach for parameterization of upright convection, leaving slantwise convection to occur explicitly on the grid." Molinari and Dudek (1992) note that option (i) is limited by a lack of data necessary to formulate a closure for a slantwise convection parameterization. Option (ii) would only succeed if the resolution were sufficient to resolve both upright and slantwise convection. [If this were the case, then the closure problem of option (i) would not be an issue.] Even with higher resolution, the unrealistic simultaneous occurrence of upright convection and slantwise convection could not always be prevented. Thus, the fully explicit approach is not optimal. Option (iii) is probably the most viable, however, only if the upright convection does not adversely influence the development of the explicit slantwise convection (Kuo and Low-Nam 1990). For the details of the parameterizations of slantwise convection, the reader is referred to Emanuel (1983), Seltzer et al. (1985), or Nordeng (1987).

d. Cumulus parameterization schemes

The main premise of the cumulus parameterization scheme is that the influence of deep convection on the environment can be accomplished using appropriate moist adiabats or one-dimensional cloud models of varying levels of sophistication to determine cloud profiles of heating and moistening through the assumption that the vertical heating and moistening on the cloud scale is the same as that on the grid scale (e.g., Kuo 1965; P84; Anthes 1977). Pielke (1984) points out, however, that this assumption is not necessarily true for shallow convection or when there are large convective downdrafts.

A useful generalized framework for cumulus schemes is described by Grell et al. (1991), who incorporate terminology proposed by Schubert (1974). The first part of this framework describes how the environment modulates convection—that is, the dynamic control or closure problem. Following Grell et al.

(1991), closure can be accomplished through 1) removal of buoyant energy in the environment during some characteristic timescale (Kreitzberg and Perky 1976, 1977; Fritsch and Chappell 1980a,b); 2) integrated moisture convergence (e.g., Kuo 1965; Anthes 1977), integrated vertical moisture advection (Molinari and Corsetti 1985), or low-level convergence (Frank 1984); or 3) an equilibrium hypothesis so that convection modifies conditionally unstable air masses toward thermodynamic equilibrium or quasi equilibrium in a specified time (e.g., Arakawa and Schubert 1974). In larger-scale models, quasi-equilibrium assumptions are often imposed. An example of a quasi-equilibrium closure is that clouds respond to changes in larger-scale circulations such that the existing level of conditional instability is maintained. In this type of closure, stabilization might be temporarily obtained by convective downdrafts cooling and drying the low levels. Convection would not occur again for several hours until the boundary layer became unstable due to sensible and latent heat fluxes or some other mechanism.

The second part of the framework describes the feedback of convection to the large scale (Grell et al. 1991), which can be depend on 1) temperature and moisture differences between cloud and environment (e.g., Kuo 1965, 1974; Anthes 1977); 2) semiarbitrary specification of vertical heating and moistening distributions (Anthes et al. 1987); 3) subsidence in the environment, entrainment at downdraft bottoms, and detrainment at updraft tops (e.g., Arakawa and Schubert 1974); and 4) lateral mixing of the cloud with the environment and subsidence (Fritsch and Chappell 1980a,b; Kreitzberg and Perkey 1976, 1977).

The last part of the framework describes the thermodynamic properties of the cumulus clouds (static control) as needed by both the closure and the feedback aspects of a cumulus parameterization (Grell et al. 1991). This can be determined by appropriate moist adiabats or by one-dimensional cloud models of varying degrees of sophistication. Some popular cumulus parameterization schemes include, as noted in part by P84, those developed by Kuo (1974), Krishnamurti and Moxim (1971), Ooyama (1971), Arakawa and Schubert (1974), Yanai et al. (1973), Betts (1975, 1986), Kreitzberg and Perky (1976, 1977), Anthes (1977), Johnson (1976), Fritsch and Chappell (1980a,b), Molinari (1982), Frank (1984), Bougeault (1985), Frank and Cohen (1987), Emanuel (1991), and others.

The simplest cumulus parameterization scheme is called the moist convective adjustment (e.g., Gerrity 1977). The premise of this scheme is that as soon as the atmosphere establishes negative vertical gradients of equivalent potential temperature ($\partial \theta_e / \partial z < 0$), it overturns; the atmospheric lapse rate is forced to be moist adiabatic in each layer or specified layers of a vertical column where saturation occurs. This approach has the desirable characteristic of being computation-ally inexpensive and simple to code (F83). In addition, there is no need to specify the physical processes responsible for convective overturning, just the end result. Limitations with the moist convective adjustment include immediate removal of grid-scale potential instability, unrealistic subgrid-scale convective heating and moistening distributions, and inaccurate precipitation rates. Finally, this scheme does not account for larger-scale forcing, which often acts to focus or trigger convection in nature. Consequently, it is very difficult to study convective-scale and mesoscale interactions with models using a moist convective adjustment scheme. In an attempt to improve on this type of scheme, Manabe et al. (1965), Miyakoda et al. (1969), and Kurihara (1973), for example, proposed the soft moist convective adjustment. The concept behind the modification is that heating and moistening should occur only on a small portion of each grid volume. This, however, did not correct many of the problems with the original scheme.

Most modern cumulus parameterization schemes use one-dimensional cloud models to determine cloud profiles of heating and moistening. The source of heating determined by these models results from latent heat release in the cloud; vertical eddy transport of sensible heating; warming (cooling) due to adiabatic subsidence (expansion); and cool, rainy, convective downdrafts. In some of these schemes (e.g., Kuo 1974), heating due to compensating subsidence is accounted for implicitly by the vertical advection term of the thermodynamic energy equation used to derive the parameterization. Moreover, it is asserted by Kuo (1974) that the net subgrid effect of convection on the resolvable scale is due to latent heat release and vertical eddy transport of sensible heat if convective drafts result in no net mass transport. In the Fritsch and Chappell (1980a,b) scheme, however, the influence of subsidence heating in a volume containing convection is computed explicitly by balancing the mass continuity equation. In contrast to many cumulus parameterizations, subgrid heating is determined with the Arakawa and Schubert (1974) scheme by the work done by a spectrum of cloud sizes (depths), in agreement with at least some observations (e.g., Lord 1982; Payne 1982).

Besides accounting for convective heating, many cumulus schemes also approximate the influences of cool low-level downdrafts (e.g., Fritsch and Chappell 1980a,b; Molinari and Corsetti 1985; Betts 1986; Frank and Cohen 1987), detailed microphysics (e.g., Kreitzberg and Perkey 1976, 1977; Fritsch and Chappell 1980a,b), and ice-phase microphysics (e.g., Zhang 1989; Zhao 1993).

The limitations of cumulus parameterization schemes have been discussed in past reviews including F83, P84, and MD92. For example, Anthes (1977) showed that results using his method were sensitive to cloud size because narrow clouds entrain more than wide clouds. It has also been shown that using one cloud size might be sufficient for small grid resolutions; however, it might be more advantageous to use a spectrum of clouds for larger grid resolutions. As another example, Rosenthal (1979) showed that his tropical cyclone model was sensitive to the vertical heating and moistening profiles produced by the one-dimensional cloud model used in his cumulus scheme. It has also been shown by various investigators that there can be sensitivities to the partitioning of moisture convergence between storage (moistening) and precipitation.

Most cumulus schemes can be classified based on their development for use in either mesoscale models or coarse grid models (F83), with the former being defined as having a grid resolution less than 50 km. For mesoscale applications, popular schemes include those developed by Kreitzberg and Perky (1976, 1977), Fritsch and Chappell (1980a,b), and Frank and Cohen (1987). For large-scale applications, popular schemes include those designed by Kuo (1965, 1974) and Arakawa and Schubert (1974), as well as various moist convective adjustments (Manabe et al. 1965; Miyakoda et al. 1969; Kurihara 1973). Some schemes developed for use in large-scale models have also been used in mesoscale models and vice versa. Examples include the schemes developed by Arakawa and Schubert (1974), Kuo (1965, 1974), and Perky and Kreitzberg (1976, 1977). An important difference between schemes appropriate for mesoscale models versus larger-scale models is that those for larger-scale models need to account for effects of mesoscale organizations that cannot be explicitly resolved (F83). In addition, cumulus parameterizations for mesoscale models need to be more accurate in specifying heating and moistening rates because heating on small dynamic scales can influence divergence on very large physical scales (F83).

4.4. Initialization and assimilation

Accurate initial conditions of the dependent variables in mesoscale numerical models including pressure, temperature, water substance, winds, and surface conditions are required to make useful and timely predictions of precipitation with mesoscale models. Unfortunately, the resolution of the operational database of the dependent variables is usually much coarser than model resolution. Therefore, we must be creative in our attempts to extract as much information as possible from the observational database.

Some of the most troublesome errors in precipitation forecasts are those associated with the initial conditions of the temperature, vapor, cloud, and precipitation fields (e.g., Smagorinsky et al. 1970; Zhao 1993). These errors may result from poor analysis and initialization techniques, lack of adequate data, insufficient resolution, and inaccurate initial conditions in the divergence and vertical motion fields.

The spinup problem of moisture and precipitation (e.g., Puri and Miller 1990; Turpeinen 1990) is prob-

ably one of the most significant problems for early forecast periods and may adversely influence subsequent forecast periods (Donner 1988; Donner and Rasch 1989; Krishnamurti et al. 1988; Krishnamurti et al. 1990). Including cloudiness—either diagnostically (Slingo 1980, 1987) or explicitly (e.g., Zhao 1993)—may help correct the spinup problem. Forecast improvements have also been made by making better estimates of the initial vertical motion, divergence, and moisture fields (Lejanas 1979; Wolcott and Warner 1981; Krishnamurti et al. 1984; Krishnamurti et al. 1988; Krishnamurti et al. 1990).

The initialization of moisture and its associated heating in mesoscale numerical models can be accomplished by either diabatic normal-mode initialization, direct specification, or thermodynamic consistency schemes, each of which is outlined below [based on information provided by Dr. Fred Carr (1992, personal communication) and Barker et al. (1994)]. The data for these different moisture initialization techniques can come from a variety of sources, including satellite, radar, and ground measurements, as summarized in Table 4.1.

With the diabatic normal-mode initialization scheme, a heating forcing term is included in the normal-mode initialization procedure. Values of this term could come from either model data (Errico and Rasch 1988) or from observations such as satellite (Turpeinen et al. 1990) and radar. As normal-mode initialization modifies the divergent wind fields, it may not be acceptable in very high resolution mesoscale models that are designed for prediction of storm-scale convection. In addition, it is difficult to obtain accurate vertical profiles of diabatic heating from available observations.

The direct specification approach assumes that observed or inferred precipitation rates can be used to infer total diabatic heating rates (e.g., Molinari 1982; Danard 1985; Benoit and Rasch 1987; Wang and Warner 1988; Carr and Baldwin 1991). Vertical heating profiles in a model are forced to be equivalent to those inferred from observed precipitation rates. This technique is generally performed during dynamic initialization or during the first few hours of the prediction

TABLE 4.1. Sources for atmospheric moisture data.

Source	Comments
Rawinsonde:	Coarse resolution
Ground observations:	No vertical resolution
Outgoing longwave radiation data:	Successful in Tropics for rainfall data
WSR-88D radar: (NEXRAD)	Surface details appear to be good Good vertical resolution is also available
Dual polarization:	Improvement in radar-derived precipitation estimates and identification of precipitation types
Satellite information:	Initialization of cloud and vapor fields

cycle using nudging. A problem with the direct specification approach is that vertical heating profiles are difficult to accurately specify for regions of precipitation not predicted by a model. A possible solution is to use the heating profile at the closest grid point in the model with precipitation or to specify the heating profile with a simple function that seems reasonable, such as a parabola.

The thermodynamic consistency approach uses the model temperature and moisture profiles, the model precipitation algorithm, and the implied heating from observed precipitation. This technique can be implemented by assuming that the model equations are imperfect and adjusting the precipitation algorithm so that the best forecast can be made (Carr and Bosart 1978). Alternatively, the precipitation algorithms can be assumed correct and the model temperature and moisture fields as modified to produce the observed precipitation (Krishnamurti et al. 1988; Donner 1988).

Another important topic in mesoscale numerical modeling involves the development of reliable initialization and assimilation procedures for surface physics data, some of which are not readily available. The requirements for initialization and assimilation of variable surface physics information includes, but is not limited to, multilevel soil moisture and vegetation type, and high-resolution surface fluxes. Some possible sources for soil moisture data might include information from satellite, rainfall histories (possibly coupled with land type and use, and vegetation), and in situ measurements. Satellite information seems ideal since it might be possible to obtain multilevel measurements for both the surface and root zone. The global vegetation index (GVI) produced weekly by the National Oceanic and Atmospheric Administration is an important source of data; however, the time between compilation of each dataset is rather long (e.g., on the order of a week). Furthermore, the resolution of this dataset is not yet suitable for mesoscale modeling. Both radar and rain gauges might also be useful in providing a history of rainfall, especially with the nationwide installation of the WSR-88D radar (NEXRAD—Next Generation Weather Radar). A recent example of global soil moisture estimates from precipitation and air temperature is described by Mintz and Walker (1993). A technique for assimilating moisture into mesoscale models is discussed by Bouttier et al. (1993a,b).

4.5. Discussion and conclusions

Based on the number and types of cumulus parameterization schemes that have been developed in the past 25 years, it does not seem likely that a scheme will be developed that is applicable for all grid resolutions. Reasons for this, as pointed out by F83, are that new observations and refinements in conceptual models typically lead to further constraints for parameterizations. At best, it is likely that schemes will be designed to perform at various dynamical and physical scales. Early examples of this trend include the development of schemes by Kreitzberg and Perky (1976, 1977) and Fritsch and Chappell (1980a,b) for simulations of mesoscale convective systems using grid resolutions of 20 km. More recently, failures with various cumulus parameterizations schemes have led to the development of hybrid schemes (e.g., Frank and Cohen 1987) for use with grid resolutions of 10–50 km.

As pointed out by MD92, one of the more difficult problems that will continue to hinder mesoscale modelers in the foreseeable future is representing precipitation processes for the grid resolutions of 3–20 km (e.g., Fig. 4.4). Unfortunately, it is not clear whether explicit or parameterization schemes should be used or if a viable approach for this scale even exists. At this scale, the largest convective storms are often marginally resolved; however, the resolution is too poor for explicit schemes to accurately represent precipitation production and the associated heat exchange with the environment, as well as the subsequent environmental response. Furthermore, most cumulus parameterizations are founded on the idea that convection is occurring in only a small portion of a grid volume, which cannot be guaranteed at these resolutions. In mesoscale systems that form in weakly forced environments with large convective instability (e.g., some squall lines, mesoscale convective systems, etc.), explicit physics may not properly model convective initiation, heat and moisture redistribution associated with convection, and the mesoscale response due to convection. It is possible that convection may actually be resolved and parameterized (double counted) simultaneously when a cumulus parameterization is used on the scale of 3–20 km, in which case the explicit convection usually dominates (MD92). A drastic alternative to modeling on the 3–20-km scale is that mesoscale modelers might consider not making predictions on this scale (MD92). Instead, they might increase the resolution in their models so that explicit precipitation physics would be more valid. However, global mesoscale numerical models are now starting to become a reality with the advent of powerful massively parallel processing computers. In the second half of the 1990s, it is quite possible that global mesoscale numerical models with resolutions of 30–50 km will become more commonplace, and by the turn of the century, it is possible that resolutions will be lowered to 15–25 km. Thus, many global mesoscale modelers will be faced with the dilemma of choosing either to model precipitation explicitly or to parameterize precipitation processes. Therefore, further efforts to resolve the problem of how to parameterize convection when grid resolution is just above that need to explicitly resolve convection are much needed.

While much worthwhile effort has been put forth

developing new cumulus parameterization schemes, the difficult task of general validation of cumulus parameterizations schemes is equally important (MD92). Attempts have been made, by doing numerous case studies, to evaluate the behaviors of cumulus parameterization schemes (Kuo and Low-Nam 1990; Anthes et al. 1989; Grell 1993) and the performance of such schemes (Zhao 1993). Cumulus parameterizations might also be evaluated for accuracy and robustness (sensitivity) by using finescale observations and/or three-dimensional, cloud-resolving numerical models. These observations and models can be used to approximate the appropriate heating and moistening of the mesoscale environment by convection, as well as the subsequent mesoscale feedbacks onto the convective scale (e.g., Soong and Ogura 1980; Benniston and Sommeria 1981; Tao et al. 1987; Krueger 1988). Use of cloud ensemble models can also provide information on the small scale that will probably not be available from observations. Another approach to validating cumulus parameterization schemes is the semiprognostic method—or one-time-step method (e.g., Lord 1982; Grell et al. 1991). As the semiprognostic method does not require the explicit integration of a numerical model, errors can be isolated from those in the observations and the cumulus parameterizations tested. An extensive review of semiprognostic tests using the Kuo (1974), Arakawa and Schubert (1974), and Kreitzberg and Perky (1976) cumulus schemes is presented by Grell et al. (1991).

In conclusion, the very nature of mesoscale models makes them sensitive to the initiation of moisture variables and representation of moisture physics. Molinari and Dudek (1992) recommend that work needs to continue to focus on addressing the validity of the parameterizations and the sensitivity of models to the parameters in these parameterizations. This work will necessarily require cloud-scale models, mesoscale models, in situ observations, remotely sensed observations (e.g., radar and satellite), and detailed case and sensitivity studies. Precipitation data of unprecedented coverage and quality will become increasingly available from the new nationwide WSR-88D network and the Oklahoma mesonet, for example, to facilitate some of these studies. In addition, data from polarimetric radars such as those in Colorado and Oklahoma (e.g., Keeler 1989; Zrnić 1991; Herzegh and Jameson 1992) might be useful. For example, Straka and Zrnić (1993) have developed an automated algorithm that uses polarimetric information to infer bulk hydrometeor types, amounts, and median sizes. For surface moisture physics and related parameterizations, higher- (time and space) resolution information of currently available parameters such as snow cover, albedo, sea temperature, land temperature, rain totals, GVI, and others could be of great use. In addition, more research is needed in retrieving the vegetative index and soil moisture because surface vegetation and surface and substrata moisture can have a strong influence in explicitly and implicitly controlling surface heat, moisture, and momentum fluxes. Finally, because of the many known sensitivities of mesoscale models to cumulus and surface physics parameterizations, interpretation of simulation results should be made with full understanding of the limitations of parameterizations (e.g., MD92).

Chapter 5

Inclusion of Radiation in Mesoscale Models

BURKHARDT ROCKEL AND EHRHARD RASCHKE

GKSS Forschungszentrum, Institut für Physik, Geesthacht, Germany

5.1. Introduction

In recent years large efforts were made to develop efficient and accurate computer codes for modeling the radiative transfer in large-scale models (GCMs and climate models). These resulted in a relatively high standard for radiative transfer models. In the many 3D mesoscale models, however, radiative energy transfer is often still treated in a less careful way. Reasons for this are, for example, that radiation as a heating or cooling source plays a more significant role in long-term climate modeling than in most smaller-timescale cases investigated by most mesoscale modeling equations. However, in the near future it will be also possible to incorporate more physical and consistent radiation parameterizations in "operational" mesoscale models.

The international scientific community successfully initiated the project ICRCCM (Intercomparison of Radiation Codes in Climate Models) to intercompare and "calibrate" the performance of the various radiation codes in use (Ellingson and Fouquart 1991).

This chapter provides only a short overview of the different basic types of radiative transfer models currently used in mesoscale modeling. It is written for mesoscale modelers who are not radiation experts but are interested in the radiation–mesoscale model interactions of radiative transfer models currently used. It does not claim to cover all radiation parameterizations. More detailed reviews of radiation codes in atmospheric models are presented, for instance, in papers by Stephens (1984) or Savijärvi (1990). In section 5.2 the interface between mesoscale models and radiative transfer modules is discussed. In the section 5.3 basic types of radiation models used in mesoscale atmospheric modeling are described. The radiative effects of clouds are summarized in section 5.4.

5.2. Radiation in mesoscale modeling

The impact of radiative energy on atmospheric processes is treated in two main components of a mesoscale model: the equations for the surface energy budget and the atmospheric temperature. Further it has some impact on the growth and decay of droplets and of ice crystals in clouds (e.g., Zhang et al. 1992). These are described briefly below.

The surface energy budget may be written in the form

$$F_h + F_w + F_r + F_c = 0, \qquad (5.1)$$

where F_h is the sensible heat, F_w is the latent heat, F_r is the radiative flux, and F_c is the conduction heat flux.

The flux of radiative energy F_r is the sum of incoming and outgoing solar and terrestrial radiative energy fluxes at the surface:

$$F_r = F_{sol}^{\downarrow}(1 - A_s) + \epsilon_s(F_{ter}^{\downarrow} - \sigma T_s^4), \qquad (5.2)$$

where $F_{sol}^{\downarrow}$ is the incident solar radiation at the earth's surface, A_s the solar albedo of the surface, $F_{ter}^{\downarrow}$ is the downwelling terrestrial atmospheric radiation, ϵ_s is the thermal surface emittance, σ is the Stefan–Boltzmann constant, and T_s is the surface temperature. All parameters in (5.2), except σ and T, are averages over their respective broad spectral ranges.

The prognostic temperature equation for the local change of the temperature may be written as

$$\frac{\partial T}{\partial t} = \text{ADV}(T) + \text{DIF}(T) + Q_l + Q_r + Q_f, \qquad (5.3)$$

where the abbreviations ADV and DIF stand for temperature changes due to advective and diffusional processes, Q_l describes latent heat changes, Q_r the radiative heating, and Q_f the heat changes due to friction.

In a cloud-free atmosphere, temperature changes in the boundary layer in mesoscale range are mainly due to advective and diffusional processes. Above the boundary layer the magnitude of temperature change decreases to about the same size as radiative heat changes (see Fig. 5.1b).

In a cloudy atmosphere (see Fig. 5.1a) latent heat changes also become important, and temperature changes due to radiative processes are large only near the top of cloud layers due to strong terrestrial cooling.

Assuming horizontal homogeneity, the radiative heating term in (5.3) is defined as

$$Q_r = -\frac{1}{\rho c_p} \frac{\partial F_r}{\partial z}, \qquad (5.4)$$

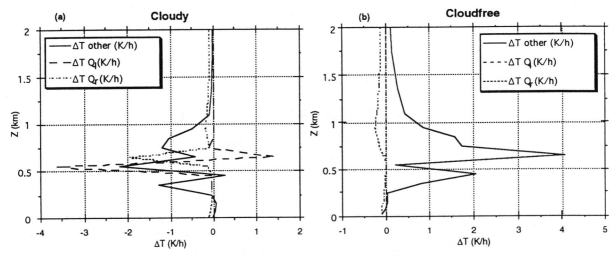

FIG. 5.1. Temperature changes due to radiation (short dashed), latent heat transformations (long dashed), and other effects (solid). Data were taken from a simulation run of the mesoscale model GESIMA (Geestacht Simulation Model of the Atmosphere) for maritime cumulus cloud fields during nighttime (i.e., terrestrial radiation only), (a) in a cloudy area and (b) in a cloud-free area.

where ρ is the mean density of air in the considered atmospheric layer, c_p is the constant of specific heat at constant pressure, z is the geometric height, and F_r is the net radiative flux at that level.

Computational methods to obtain the components of F_r will be subject of considerations in section 5.3.

The direct impact of radiation on the growth of cloud particles (cloud droplets or ice crystals) is normally negligible in mesoscale forecast models. It may be considered for mesoscale research models for special cloud case studies, where a deeper insight into the cloud mechanisms and life cycle is the main objective. Size, form, and lifetime of the crystals will be influenced by the direct impact of radiation. Studies concerning the influence of radiation on the growth cloud particles were carried out by, for example, Zhang et al. (1992) and Stephens (1983).

The prognostic mass equation of a cloud droplet can be written in the form

$$\frac{\partial m}{\partial t} = 4\pi CD(\rho_V - \rho_{V_d}) - \frac{A}{L}F_r, \qquad (5.5)$$

where m is the mass of a droplet and C is its "capacitance" describing the form of the particle (for spheres C is the radius), D is the molecular diffusion coefficient, ρ_V is the ambient vapor density, and ρ_{V_d} is the vapor density of the droplet surface. The second term on the right side describes the changes of droplet mass due to absorption of radiative energy. Here, A is the cross-sectional area of the droplet and L is the heating due to sublimation or condensation.

5.3. Radiation models (clear sky)

We can distinguish between the three following major groups of radiation models used in the mesoscale

domain: surface radiation budget models, heating–cooling rate models, and flux density models.

In this section we will give some examples for these models. We do not claim to present a complete synopsis.

a. Surface radiation budget models

The up- and downward directed fluxes over the entire solar and terrestrial spectrum can, in general, be described with the following generalized form:

$$F_{\text{sol}}^{\text{net}} = S_0 \cos\Theta f(\tau_1, \cdots, \tau_n, A_S)$$

$$F_{\text{ter}}^{\downarrow} = \sigma T_a^4 f(\epsilon_1, \cdots, \epsilon_n)$$

$$F_{\text{ter}}^{\uparrow} = \sigma T_s^4 \epsilon_s, \qquad (5.6)$$

where $F_{\text{sol}}^{\text{net}}$ is the net solar flux, $F_{\text{ter}}^{\downarrow}$ is the downcoming terrestrial flux, and $F_{\text{ter}}^{\uparrow}$ is the flux emitted by the ground. Here, S_0 is the solar constant, Θ is the solar zenith angle, A_S is the solar surface albedo, σ_B is the Stefan–Boltzmann constant, T_a is the atmospheric temperature, T_S is the surface temperature, and ϵ_S is the surface emittance. Here, $\tau_1, \ldots \tau_n$ are empirical transmission functions for molecular absorption of different absorbers, scattering, aerosol, and cloud effects; $\epsilon_1, \ldots, \epsilon_n$ are empirical emission functions for different absorbers.

Mahrer and Pielke (1977) introduced the following relationship for the net solar flux at the ground:

$$F_{\text{sol}}^{\text{net}} = S_0 \cos\Theta(1 - A_S)(\tau_E - a_V), \qquad (5.7)$$

where τ_E is an empirical transmission function after Kondratyev (1969) and Atwater and Brown (1974) accounting for molecular scattering and absorption by oxygen, ozone, and carbon dioxide, as well as for the forward Rayleigh scattering effects:

$$\tau_E = 0.485 + 0.515$$

$$\times \left[1.041 - 0.16 \left(\frac{0.000949 p + 0.051}{\cos \Theta} \right)^{1/2} \right], \quad (5.8)$$

where p (hPa) is atmospheric pressure. Here, a_V describes the absorption by water vapor after McDonald (1960):

$$a_V = 0.077 \left[\frac{u(z)}{\cos \Theta} \right]^{0.3}, \quad (5.9)$$

where $u(z)$ is the total water vapor content of the atmosphere above the layer z.

The downward atmospheric heat radiation is described by Mahrer and Pielke (1977) by a sum over the contributions of different atmospheric layers to the total downcoming terrestrial radiation at the ground:

$$F_{ter}^{\downarrow} = \sum_{j=1}^{top-1} \frac{\sigma}{2} (T_{j+1}^4 + T_j^4)[\epsilon(1, j+1) - \epsilon(1, j)]$$

$$+ \sigma T_{top}^4 [1 - \epsilon(1, top)]. \quad (5.10)$$

Since in this equation the atmosphere is not treated as an entity but rather is divided into single layers with different emittances, this is not a parameterization of the surface flux as defined by (5.6) where the atmosphere is treated as a single layer. It is more a simplified part of a flux density model. Mahrer and Pielke used empirical functions of ϵ for water vapor and carbon dioxide emittances.

Somieski et al. (1988) introduced aerosol absorption and scattering in a modified and extended version of the Mahrer and Pielke model.

The net solar flux at the ground is described by

$$F_{sol}^{net} = S_0 \cos \Theta \tau_E \tau_V \tau_{AA} \tau_{AS}[1 + 0.1(A_S - 0.2)], \quad (5.11)$$

where τ_E is the transmission function as defined by (5.8) for Rayleigh scattering and absorption by permanent gases; τ_V, τ_{AA}, and τ_{AS} are the transmission functions for absorption by water vapor and absorption and scattering by aerosols, respectively. Empirical formulas are used for τ_E, τ_V, τ_{AA}, and τ_{AS}. Absorption by water vapor is slightly different from that used by Mahrer and Pielke (1977):

$$\tau_V = 1 - 0.093 \left(\frac{u}{\cos \Theta} \right)^{0.37}. \quad (5.12)$$

The parameterization of aerosol influence on the solar surface flux is derived from Atwater and Brown's (1974) formulation

$$\tau_{AA} = \exp \left[- \frac{0.114}{\cos \Theta} \int_{surface}^{top} \sigma_e(0.55) dz \right] \quad (5.13)$$

and

$$\tau_{AS} = 0.97 \tau_{AA}, \quad (5.14)$$

where $\sigma_e(0.55)$ is the aerosol extinction coefficient at 0.55 μm.

The Ångström formula (Linke 1970) is used by Somieski et al. (1988) to derive the downward terrestrial flux density

$$F_{ter}^{\downarrow} = \sigma_B T_a^4[0.79 - 0.174 \exp(-0.095 e_a)], \quad (5.15)$$

where T_a is the atmospheric temperature and e_a (hPa) the water vapor pressure, both 2 m above ground.

A more recently published parameterization of surface radiation budget is described by Savijärvi (1990). The net solar radiation at ground is calculated by

$$F_{sol}^{net} = S_0 \cos \Theta [1 - \alpha_o - a_a \alpha_w - a_s(\alpha_R - 0.07 A_S)], \quad (5.16)$$

where a_a and a_s are coefficients describing the aerosol absorption and scattering, respectively. Savijärvi suggested from observation values of $a_a = 1$ and $a_s = 1.9$ for atmospheres with small and nonabsorbing aerosol particles (e.g., over oceans) and values of $a_a = 1.2$ and $a_s = 1.25$ over continental industrialized areas. In principle, values of a_a and a_s should be determined empirically from measurements. The coefficients α_o, α_w, and α_R describe the absorption by ozone, water vapor (plus overlapping effects by carbon dioxide and oxygen), and the Rayleigh scattering, respectively:

$$\alpha_o = 0.024 (\cos \Theta)^{-0.5}$$

$$\alpha_w = 0.11 \left(\frac{\tilde{u}}{\cos \Theta} \right)^{0.5}$$

$$\alpha_R = \frac{0.28}{1 + 6.43 \cos \Theta}, \quad (5.17)$$

where $\tilde{u}$ is the pressure-scaled water vapor amount of the total atmosphere as defined by

$$\bar{u} = \int_0^{p_s} q \left(\frac{p}{p_0} \right)^{0.85} \left(\frac{T_0}{T} \right)^{0.5} \frac{dp}{g}, \quad (5.18)$$

where q is the specific humidity, g is the gravity, $p_0 = 1013$ hPa, and $T_0 = 273$ K.

Results of tests with this parameterization for areas in eastern Australia and southern Finland are published by Savijärvi (1990).

For the terrestrial downward radiation Savijärvi (1990) considers an integration of the downward flux over the whole atmosphere for the absorption by water vapor rotation bands. Continuum absorption is mainly effective in the layers near ground; thus the approximate flux is a function of the specific humidity at the ground. A constant flux of 16 W m^{-2} is assumed for carbon dioxide, ozone, and aerosols together. Savijärvi combines this to the following simple equation for the downward terrestrial radiation at the ground:

$$F_{ter}^{\downarrow} = F_w^{\downarrow} + 4 q_s + 16, \quad (5.19)$$

where $F_w^\downarrow$ is the integrated flux due to water vapor rotation bands and q_s (g kg^{-1}) is the specific humidity at the ground.

b. Heating and cooling rate models

These models provide directly heating and cooling rates for each atmospheric layer of the mesoscale model. One can divide these models into two groups: the first group treats the computation of heating and cooling rates similar to the flux calculation in the surface radiation budget models—that is, by transmission functions. These types of models are mainly used in the solar region. The second group is derived from explicit flux density models and are mainly used for determining terrestrial cooling effects.

Somieski et al. (1988) presented an equation for the solar heating in the form

$$\left(\frac{\partial T}{\partial t}\right)_{sol} = \frac{1}{\bar{\rho} c_p} S_0 f_s \tau_E \tau_{AA} \tau_{AS} [1 + 0.1(A_S - 0.2)]$$

$$\times \left[0.114 \tau_w \sigma_e(0.55) - 0.0344 \left(\frac{u}{\cos\Theta}\right)^{-0.63} \frac{\partial u}{\partial z}\right],$$

(5.20)

where f_s is the correcting factor for the influence of surface reflection on the upward solar flux, ρ is the density of the air, and c_p is the specific heat constant; all other parameters are as defined in (5.11).

Savijärvi (1990) introduced a solar heating rate with two terms, one representing the variable influence of water vapor amount on the solar heating rate, the other assuming a constant heating rate for carbon dioxide, oxygen, and ozone absorption together:

$$\left(\frac{\partial T}{\partial t}\right)_{sol} = S_0 \left(\frac{q}{c_p}\right) \left(\frac{p}{p_0}\right) \left\{y\left[\frac{\tilde{u}(z, \infty)}{\cos\Theta}\right]\right.$$

$$+ 1.67 A_S \cos\Theta y \left[\frac{\tilde{u}(0, \infty)}{\cos\Theta} + \tilde{u}(0, z) 1.67\right]\right\}$$

$$+ 1.7 \times 10^{-6} (\cos\Theta)^{0.3} \quad (5.21)$$

with

$$y(\bar{u}) = \begin{cases} 0.029 \bar{u}^{-0.81}, & \bar{u} \geq 0.05 \text{ cm} \\ 0.050 \bar{u}^{-0.63}, & \bar{u} < 0.05 \text{ cm}, \end{cases} \quad (5.22)$$

where q is the specific humidity. The other variables are as previously explained for (5.16) and (5.17). Aerosol effects are neglected in this formulation but may be crudely included by increasing the constant 0.029 in (5.22) to 0.04–0.05 near the surface, or again empirical regressions could be made.

Two cooling rate approximations for terrestrial radiation may be mentioned here. The first one is the so-called "cooling to space" approximation by Rodgers and Walshaw (1966) and Rodgers (1967), which can

be derived from combining their emissivity flux equations into the general cooling rate equation

$$\left(\frac{\partial T}{\partial t}\right)_{ter} = \frac{g}{c_p} \frac{\partial(F_{ter}^\uparrow - F_{ter}^\downarrow)}{\partial p} \quad (5.23)$$

and neglecting all terms but that one representing the emission of atmosphere from level z to space [$\epsilon(z, \infty)$]:

$$\left(\frac{\partial T}{\partial t}\right)_{ter} = \frac{g}{c_p} \frac{\partial \epsilon(z, \infty)}{\partial p} \sigma_B T^4(p). \quad (5.24)$$

The second one can be derived by the flux formulation published by Sasamori (1972) where the atmosphere is assumed to be isothermal at the level of interest:

$$\left(\frac{\partial T}{\partial t}\right)_{ter} = -\frac{\sigma_B}{\bar{\rho} c_p} \left\{\frac{\epsilon(0, z)}{\partial z} [T^4(z) - T_s^4]\right.$$

$$\left. - \frac{\partial \epsilon(z, \infty)}{\partial z} [T_{top}^4 - T^4(z)]\right\}. \quad (5.25)$$

c. Flux density models

A more consistent way to consider radiation in mesoscale modeling is the use of radiative flux density models. These models provide radiative fluxes at each level in the atmosphere. From these fluxes the heating or cooling rate for each layer can be computed by (5.4). Thus both the surface radiation budget and the heating or cooling rate of the atmospheric model layers are determined by a single model. The flux density models provide more information than the two other models together. The computed fluxes may be used for instance for validation studies with satellite data.

Under the aspect of an optimal combination of accuracy and low computational time, another type of flux density model has been established during the last years: the two-stream approximated model. Within the two-stream solution of the radiative transfer equation the angular dependence of the radiation is approximated by integrating fluxes over the upward and downward half-hemispheres. For the general differential form for the two-stream approximated radiative transfer equation one obtains

$$\frac{dF^\uparrow}{d\delta} = \alpha_1 F^\uparrow - \alpha_2 F^\downarrow - \alpha_3 J$$

$$\frac{dF^\downarrow}{d\delta} = \alpha_2 F^\uparrow - \alpha_1 F^\downarrow + \alpha_4 J \quad (5.26)$$

with

$$\alpha_1 = \frac{1}{\bar{\mu}} [1 - \tilde{\omega}(1 - \beta)]$$

$$\alpha_2 = \frac{\tilde{\omega}\beta}{\bar{\mu}}$$

$$\alpha_3 = \begin{cases} \beta_0, & \text{for solar radiation} \\ \dfrac{1 - \tilde{\omega}}{\bar{\mu}}, & \text{for terrestrial radiation} \end{cases}$$

$$\alpha_4 = \begin{cases} 1 - \beta_0, & \text{for solar radiation} \\ \dfrac{1 - \tilde{\omega}}{\bar{\mu}}, & \text{for terrestrial radiation} \end{cases}$$

$$J = \begin{cases} \tilde{\omega} S_0 \exp\left(-\dfrac{\delta}{\cos\Theta}\right), & \text{for solar radiation} \\ B, & \text{for terrestrial radiation,} \end{cases}$$

where δ is the optical thickness, β_0 is the backscattering coefficients for the direct and β for the diffuse radiation, $\tilde{\omega}$ is the single scattering albedo, $1/\bar{\mu}$ is the diffusivity factor, Θ is the solar zenith angle, S_0 is the incoming solar flux at the top of the atmosphere, and B is the Planck function. All these quantities are usually taken as spectrally weighted averages over broad spectral intervals to save computational efforts. Different ways of treating the absorption of gases are shown, for example, by Stephens (1984), Lenoble (1985), Goody and Yung (1989), and Kyle (1991). Numerous approximations exist for solving the differential equations (5.26) to get efficient radiative computer codes. It is beyond the scope of this paper to explain them in detail. For further reading we refer to reviews and to the theoretical framework in the literature (see, e.g., Ritter and Geleyn 1992; Lenoble 1985; Liou 1980; Meador and Weaver 1980; Stephens 1984; Zdunkowski et al. 1980).

The two-stream approximated radiation models describe only the vertical transport of radiation in the atmosphere. This is sufficient for most applications. For studies of mesoscale atmospheric phenomena where the horizontal exchange of radiation is important (e.g., contrails) the more complex radiative transfer models must be applied. Six-stream approximations may then be used (see, e.g., Gierens 1993). Since these models are very time consuming they may not be applied to the whole atmosphere but to critical regions only. For a good representation of the rest of the atmosphere, these models can be combined with two-stream models.

d. Accuracy of radiation models

The results of the ICRCCM program have become an internationally accepted benchmark test for atmospheric models. The ICRCCM management provides input for radiation codes describing different atmospheric conditions (i.e., vertical temperature and absorber profiles, cloud, and aerosol types). A new radiation code can then be tested against the results of up to about 40 different radiation codes from spectral high-resolution to broadband models. Several research results of the ICRCCM are described in a special issue of the *Journal of Geophysical Research* [1991, **96**(D5)]. In the first test phase many of the participating radiative transfer models show unacceptable errors in some of their results. These errors could be identified and corrected in the most of the models. Ellingson et al. (1991) and Fouquart et al. (1991) summarized the longwave and shortwave results of the ICRCCM. Some of the error values are listed below.

Longwave
- Clear-sky line-by-line model (LBL) results agree within 1%. However, they should not be used as absolute reference due to uncertainties about line shape and absorption continua.
- Clear-sky medians of band model fluxes and cooling rates agree within 1%–2% with LBL results. The rms error is 5%–10%.
- For near-black clouds the results agree closely near the cloud boundaries. For optically thin clouds the spread was large (35–80 W m^{-2}).

Solar
- The rms error for clear sky conditions is about 4% for downward flux at the surface and 6%–11% for the total atmospheric absorption.
- In contrast to the longwave cases, clouds with higher optical thickness show the worst results. The rms error lies between 4% and 10%.
- The errors for aerosol cases are 1%–21% with largest errors for high aerosol concentrations.

Even if a radiative transfer routine is not flexible enough to calculate all the test cases we strongly recommend that results be compared for those test cases it is able to run. This helps to identify possible error sources and therefore may improve the accuracy. The ICRCCM input data for these tests are available on a PC compatible disk.

5.4. Radiative effects of clouds

Clouds are one of the most effective atmospheric component in modifying radiative transfer in the earth's atmosphere. More details may be found, for example, in Hansen and Travis (1974), Hobbs and Deepak (1981), and Kondratyev (1969) and in all publications on the effect of clouds on the planetary radiation budgets and elsewhere.

The way that radiative cloud properties are parameterized depends on the information about the cloud provided by the mesoscale cloud model. Three different parameterizations will be mentioned in the following. The first one is especially useful for the surface energy budget models and the other two for flux density models. All these parameterization assume vertical homogeneous cloud layers.

a. Prescribed optical cloud properties

In some mesoscale models radiative properties of clouds are prescribed. Variations in the surface energy budget due to special cloud types may be described by transmission functions for the solar radiation and an enhancement function for the terrestrial radiation. These functions can be implemented into equations of the form like (5.6). As an example, the cloud parameterization used by Anthes et al. (1987) is shown here. They define transmittances

$$\tau_{ac} = \prod_{i=1}^{3} [1 - (1 - \tau_{ai})]n_i, \quad \text{for absorption}$$

$$\tau_{sc} = \prod_{i=1}^{3} [1 - (1 - \tau_{si})]n_i, \quad \text{for scattering.} \quad (5.27)$$

Different values are assumed for high, middle, and low clouds (index $i = 1, 2, 3$) whose fractional cover is given by the number n_i.

The downward terrestrial radiation is enhanced as follows:

$$\tilde{F}^{\downarrow}_{\text{ter}} = F^{\downarrow}_{\text{ter}}\left(1 + \sum_{i=1}^{3} c_i n_i\right), \quad (5.28)$$

where c_i is the enhancement coefficient, which seems to depend a bit on the cloud-bottom temperature. Some values used by Anthes et al. are listed in Table 5.1.

b. Cloud water mass mixing ratio

In more advanced cloud schemes for mesoscale models the cloud mass mixing ratio q_w (kg kg^{-1}) of a cloud layer is a prognostic variable. From the cloud mass mixing ratio, the water content m_w (g m^{-3}) of a cloud can easily be computed by

$$m_w = \rho_a q_w, \quad (5.29)$$

where ρ_a is the density of the air.

As shown by Fouquart et al. (1990), the optical depth of a cloud for solar radiation is a function of the cloud liquid water content m_w and the effective radius r_e of the cloud drop size distribution:

$$\delta_c = \frac{3m_w \Delta z}{2\rho_w r_e}, \quad \text{for} \quad r \gg \lambda, \quad (5.30)$$

where Δz is the geometrical thickness of the cloud layer and ρ_w the density of the water droplet.

The effective radius is not determined by the cloud model, but for the "standard clouds" published by Stephens (1978, 1979) the effective radius might be approximated by a linear function of the liquid water content:

$$r_e = 11m_w + 4, \quad (5.31)$$

with r_e in microns and m_w in grams per cubic meter.

Besides the optical thickness, the asymmetry factor g_c (accounting for the asymmetry of the scattering function) and the single scattering albedo $\tilde{\omega}_c$ (determining the ratio of cloud scattering to extinction) de-

TABLE 5.1. Transmittances τ and enhancement coefficients c on radiative flux modification due to clouds (from Anthes et al. 1987).

Cloud level	τ_{ai}	τ_{si}	c_i
High	0.89	0.80	0.06
Middle	0.85	0.60	0.22
Low	0.80	0.48	0.26

scribe the optical properties of clouds. For g_c a mean value 0.85 might be assumed. The single scattering albedo is parameterized by Fouquart et al. (1990) as a function of the solar zenith angle Θ, and the optical thickness is computed with (5.30):

$$\tilde{\omega}_c = 1 - 10^{-3}[0.9 + 2.75(\cos\Theta + 1)\exp(-0.09\delta_c)]. \quad (5.32)$$

For the terrestrial radiation, scattering is commonly neglected and a diffuse emittance ϵ_c in the form

$$\epsilon_c = 1 - \exp\left(-\frac{km_w \Delta z}{\bar{\mu}}\right) \quad (5.33)$$

is used. Here, k (m^2 g^{-1}) is the mass absorption coefficient. For "standard clouds" Stephens (1979) distinguishes between emittances for the downward ($k = 0.95$ m^2 g^{-1}) and for the upward radiation flux ($k = 0.78$ m^2 g^{-1}). Here, m_w is again the liquid water content.

c. Cloud mass mixing ratio and total number concentration

In other mesoscale research models clouds are treated in a relatively complex way. A couple of prognostic equations are built into these models describing the cloud mass of different phases of water and also the total number concentration of cloud particles. For mesoscale models running in operational mode—for example, at weather service centers—these kinds of models are still too time consuming. They restrict to prognostic equations for water vapor and liquid water only. Detailed process studies (e.g., cloud life cycle simulations) with research models, however, take advantage of these more complex cloud models.

With the total number concentration N as an additional variable, the radiative properties of clouds can be described more precisely as pointed out by Rockel et al. (1991). Radiative cloud properties in the solar region may be described as follows:

$$g_{c_x} = a_{g_x} + \log\left(\frac{m_x}{N_x}\right)^{b_{g_x}}$$

$$\sigma_{s_{c_x}} = a_{s_x}\left(\frac{m_x}{N_x}\right)^{b_{s_x}} N_x$$

$$\sigma_{a_{c_x}} = a_{a_x}\left(\frac{m_x}{N_x}\right)^{b_{a_x}} N_x, \quad (5.34)$$

where g_c, σ_{s_c}, and σ_{a_c} are the asymmetry factor, the scattering coefficient, and the absorption coefficient, respectively. Here, m (g m^{-3}) is the cloud mass, a and b are empirical constants, and the index x denotes the different water phases.

The terrestrial diffuse emittance ϵ_c can be expressed as

$$\epsilon_{c_x} = 1 - \exp\left[a_{\epsilon_x} \left(\frac{m_x}{N_x} \right)^{b_{\epsilon_x}} \frac{N_x \Delta z}{\bar{\mu}} \right], \quad (5.35)$$

where Δz is the geometrical thickness of the cloud layer and $1/\bar{\mu}$ is the diffusivity factor (commonly assumed to be 1.66).

The constants in (5.34) and (5.35) are derived as follows. For 32 water and 40 ice standard-size droplet distributions with different effective radii (from 0.2 to 80 μm), effective variances (from 0.01 to 0.25), and number concentrations (10–500 cm^{-1} for water, 0.01–1.0 cm^{-1}). Mie calculations for the solar and terrestrial spectrum were carried out. The constants were then obtained by a curve fitting. Results of other authors (e.g., Stephens 1978) where the optical parameters are functions of m_w fit well as a "mean" curve through the data from Rockel et al. (1991). However, these data are derived assuming that ice cloud particles are spherical. The asymmetry factor for irregular-shaped ice crystals is about 0.9 times less than that for spheres.

5.5. Conclusions

Of the radiation models described in this paper, the radiative flux density models provide the most physically based and consistent method for modeling the radiative transfer in the atmosphere. The main argument against the use of flux density models is the larger amount of computation time they require. However, computer capacity and speed has reached a level high enough to introduce flux density models into mesoscale modeling. The two-stream approximated radiative transfer models are used in large-scale atmospheric models as general circulation and climate models and in numerical weather prediction models. Therefore, they provide consistent parameterization from large scale to mesoscale when models with different resolutions are nested.

Chapter 6

Mesoscale Meteorological Model Evaluation Techniques with Emphasis on Needs of Air Quality Models

STEVEN R. HANNA

Sigma Research Corporation, Concord, Massachusetts

6.1. Objectives

The objective of this article is to review techniques used to evaluate mesoscale meteorological models with emphasis on the parameters used as input to air quality models. Our definition of "mesoscale" is broad enough to encompass so-called regional models. The characteristics of existing model evaluations are compared with the components of an optimum comprehensive model evaluation system.

Several examples are given of evaluations of current models that emphasize wind field pattern comparisons and trajectory comparisons. Recommendations are given for future work in this area, including performance measures and datasets.

6.2. Brief overview of previous meteorological model evaluations

As Brier (1990) points out in his review of the history of model evaluations in meteorology, up until the 1950s there were few quantitative methods for evaluating meteorological predictions. Then, as a result of controversies regarding weather modification, the pendulum swung the other way, to the extent that statisticians were planning and conducting cloud-seeding experiments (Gabriel 1987). Current model evaluation methods in use by the National Oceanic and Atmospheric Administration employ statistical procedures that would be of use to air quality modelers, but they emphasize parameters (e.g., rainfall and 500-mb heights) that are of great interest to forecasters but are of less interest to mesoscale air quality modelers (Kalnay and Dalcher 1987; Murphy and Winkler 1987; Murphy and Epstein 1989). Mesoscale and regional air quality models such as the Regional Acid Deposition Model (RADM, Chang et al. 1987) and the Acid Deposition and Oxidant Model (ADOM, Venkatram et al. 1988) are dependent on accurate values of the following meteorological input parameters: boundary layer wind, turbulence, and temperature profiles; surface momentum and heat fluxes; mean vertical motions and cloud characteristics in the lowest 2 or 3 km; and

radiation. These parameters must be supplied on a three-dimensional grid and at a time average consistent with the air quality model (usually horizontal grid distances range from about 0.5 to about 80 km, vertical grid distances range from about 10 to 1000 m, and time averages vary from about 1 min to 1 h).

Of all the parameters listed above, the boundary layer wind fields have been the primary subject of model evaluation exercises. Because models produce wind vectors on spatial grids, these wind speeds and directions can be qualitatively or quantitatively compared with observations (e.g., Kessler et al. 1988; Alpert and Getenio 1988a,b). The gridded wind fields can be used to estimate particle trajectories, which can also be compared with observations (e.g., Reisinger and Mueller 1983) or with the predictions of alternate models (e.g., Kahl and Samson 1986).

The components of the meteorological model can be *implicitly* evaluated by comparing tracer concentration distributions predicted by the air quality model with field observations (e.g., Yamada et al. 1992; Klug et al. 1992; Dennis et al. 1989). However, with such an evaluation, it is impossible to know whether or not individual scientific algorithms in the model are valid. For example, errors in algorithms could cancel each other, leading the analyst to the mistaken conclusion that all of the scientific components of the model are performing properly.

6.3. An optimum comprehensive model evaluation system

General aspects of the evaluation of environmental models are discussed in many references, including Fox (1981), Willmott (1982), Pielke (1984), Beck (1987), Hanna (1988), Venkatram (1988), Tesche (1988), and Cox and Tikvart (1990). These documents stress the need to include both the scientific and statistical components of the model evaluation study, as discussed in detail below.

a. Scientific component

The scientific evaluation of any model should be done by an expert familiar with the technical issues.

Pielke (1984) discusses the general components of a scientific evaluation. The topics listed below include several from his list, plus others suggested by Beck (1987).

(i) Peer review—The technical documentation for the model should be peer reviewed by independent experts at the level of a journal article. Technical justification for all model components should be given, including full sets of equations and assumptions.

(ii) Comparison with analytical solutions—Many numerical problems have analytical solutions for certain simplified scenarios. The numerical model predictions should agree with these analytical solutions. For example, Durran and Klemp (1983) compare the predictions of their mesoscale model with the solution for linearized inviscid hydrostatic flow over a two-dimensional bell-shaped mountain.

(iii) Diagnostic evaluation of components—any large comprehensive model consists of a set of interconnected numerical modules or components. The predictions of these components (e.g., plume-rise module, advection module, or surface heat budget module) should be compared with observations, where available. For example, Mahrer and Pielke (1978) present the results of tests of an advection scheme.

(iv) Conservation of mass and energy—Most large models contain approximations that ultimately lead to monotonic increases or decreases in the mass and/or energy integrated over the entire domain when the model is run for several days. Of course, over periods of a few hours, local forcing functions such as land–sea temperature differences can lead to increases in kinetic energy, but these variations should be cyclical.

(v) Code verity—The code should be checked line by line to verify that it is consistent with the technical documentation. This is seldom done because the codes are so long and complicated, having evolved usually over many years under the hand of many graduate students.

(vi) "Look and see"—As suggested by Mosteller and Tukey (1977), intuition should be applied to tables and plots of model output to determine if they "look" correct. For example, do the time and space patterns of drainage flows conform to what is known from observations? The human eye is a very efficient integrator of spatial information. A good example of use of this procedure is found in the report by Kessler et al. (1988) and the overview paper by Douglas and Kessler (1991), where numerous maps of wind vectors are presented and qualitatively discussed.

b. Statistical component

It is important, before one starts the statistical component of a model evaluation exercise, to first precisely define the purpose of the study. Statisticians (e.g., Panofsky and Brier 1968; Mosteller and Tukey 1977) ask that a specific hypothesis be made. For example, the

following hypothesis may be appropriate: "There is no significant difference (at the 95% confidence level) between the observed and predicted hourly averaged wind speeds for a specific hour over a network of 20 monitoring stations at a specific field site." Another example of a hypothesis might be "Model A provides a better representation (at the 95% confidence level) of a set of ten observed constant-level balloon trajectories than models B and C." The subsequent analysis is directed toward testing the hypothesis. If the hypothesis is not well defined, the study may become diffused into a set of ill-defined and unconnected statistical calculations and discussions.

The statistical model evaluation exercise can include both comparisons with other models and comparisons with field observations. Each of these analyses is based on the assumption that the various sets of data points being analyzed (e.g., data from sets of monitors in a mesoscale network or data from a time series observed at a fixed point) are independent. If not, then the so-called degrees of freedom used in the statistical tests must be adjusted downward to reflect the actual numbers of independent points in the dataset. As discussed later, the concept of independence becomes fuzzy in the case of models employing "data assimilation," where observations are used to "nudge the model" and these same observations are used later to evaluate the model.

Fox (1981) suggested a long list of statistical tests for evaluating air quality models. These tests were applied by the Environmental Protection Agency (EPA) to several scenarios, with the result that it was recommended that the list be pared down to only three or four items. Cox and Tikvart (1990) discuss some aspects of the improved statistical tests now used by the EPA. Hanna's (1989) list of model performance measures includes the relative mean bias, FB; the relative scatter, NMSE; the correlation, R; and the fraction within a factor of 2, FAC2, as defined below:

relative or fractional mean bias:

$$FB = \frac{(\bar{X}_o - \bar{X}_p)}{0.5(\bar{X}_o + \bar{X}_p)}, \qquad (6.1)$$

relative or fractional mean-square error:

$$NMSE = \frac{\overline{(X_o - X_p)^2}}{\bar{X}_o \bar{X}_p}, \qquad (6.2)$$

correlation coefficient:

$$R = \frac{\overline{(X_o - \bar{X}_o)(X_p - \bar{X}_p)}}{\sigma_{X_o}\sigma_{X_p}}, \qquad (6.3)$$

fraction of cases, FAC2, where predictions are within a factor of 2 of the observations:

$$(0.5 < X_o/X_p < 2), \qquad (6.4)$$

where X_o is an observed variable, X_p is a predicted variable, and overbars represent averages over all the data.

Note that because the means $\bar{X}_o$ and $\bar{X}_p$ are used in the denominators of the FB and NMSE expressions, these formulas are only appropriate for positive definite variables with a zero background. Ninety-five percent confidence limits on these performance measures for a single model or on the difference between these performance measures for two models can be calculated by means of Student's t-test, where the standard error can be estimated from the data using a standard resampling procedure [e.g., so-called bootstrap resampling is used by Hanna (1989)].

A similar set of statistical tests was suggested by Willmott (1982) and has been used in several mesoscale model evaluation exercises (e.g., Steyn and McKendry 1988). Willmott breaks down the mean-square error (MSE) into so-called systematic and unsystematic components, MSE_s and MSE_u, respectively:

$$MSE_s = \overline{(X_o - X_p^*)^2}, \qquad (6.5)$$

$$MSE_u = \overline{(X_p - X_p^*)^2}, \qquad (6.6)$$

where the line $X_p^* = a + bX_o$ is the result of an ordinary least-squares linear regression between X_o and X_p (note that $MSE = MSE_s + MSE_u$). Willmot defines an index of agreement d by the formula

$$d = 1 - \frac{MSE}{(X_p - \bar{X}_o)^2 + (X_o - \bar{X}_o)^2}. \qquad (6.7)$$

This dimensionless index has a theoretical range of 1.0 (for perfect agreement) to 0.0 (for no agreement).

The performance measures in (6.1)–(6.7) are more appropriate when simple sets of pairs of numbers are being analyzed, such as observed and predicted daily maximum temperatures at a given point for all July days over a 10-yr period. However, when spatial patterns are being analyzed, which are of interest in many mesoscale meteorological model applications, it is necessary to consider alternate performance measures. Klug et al. (1992) propose a "figure of merit" defined in space or time. For example, when the areas of observed and predicted precipitation are drawn on a map, then the space figure of merit is defined as the area of the intersection of the observed and predicted areas, divided by their sum (or union, in the terms of set theory). To better account for the uncertainties in spatial patterns inferred from limited monitoring networks, Dennis and Seilkop (1986) suggest the application of an interpolation procedure known as kriging. The evaluation of spatial patterns has been of special interest to regional air quality modelers, due to the often contorted patterns of the observed and predicted distributions. General discussions of evaluations of the EPA's RADM are found in papers by Barchet (1987), Barchet and Dennis (1989), and Dennis et al. (1989).

Forecasters have long used contingency tables to analyze the accuracy of their forecasts. To develop these tables, it is necessary to define certain categories (e.g., rain or no rain) and tally the numbers of concurrent observations and predictions in each intersection of categories. For example, in Table 6.1, it is seen that there were six successful predictions of rain, but there were 15 occasions when rain was observed and not predicted. The significance of the results can be assessed using standard statistical tests (Panofsky and Brier 1968). This method is useful only if the data lend themselves to categorization.

Sensitivity studies can also be thought of as statistical tests, especially when a carefully designed Monte Carlo procedure is followed (Beck 1987). In this procedure, a model is run hundreds of times. Each time a new value of one or more input parameters is selected by random sampling, and the variations of the model output are noted. The probability distribution function of input parameters is usually defined by a mean, a standard deviation, and a range. Sometimes known correlations among input parameters can be preserved in the Monte Carlo process. Because many advanced models require long run times, it is often impractical to conduct hundreds of runs with these models. Consequently, new procedures for resampling have been developed that are designed to reduce the number of runs, while still reproducing the original probability distribution function. Tesche et al. (1990) recommended that for complex regional models the number of sensitivity studies be further reduced to one or two for each input variable defined by the expected limits to the ranges of the variables.

6.4. Examples of recent model evaluations

Evaluations of mesoscale and regional-scale meteorological models have emphasized trajectories and wind fields. There have been a few studies in which boundary layer parameters such as surface heat and momentum fluxes have been evaluated. Some examples of these types of evaluations, as well as examples of comparisons of predicted and observed regional concentration patterns are given below.

a. Evaluations of mesoscale and regional trajectories

Over 20 reports and journal articles on the subject of mesoscale and regional-scale trajectory evaluation were reviewed and the results are summarized below. This type of study is so popular because 1) trajectories are obviously important in regional air pollution prob-

TABLE 6.1. Contingency table for prediction of rain.

	Predicted	
	No rain	Rain
Observed		
No rain	25	2
Rain	15	6

lems since they allow the time and location of the pollutant impact to be estimated, and 2) trajectories are relatively easy to observe with balloons released during field experiments and to calculate using modeled wind fields.

Field observations of trajectories over travel distances ranging from 10 to 10 000 km have often made use of constant-level balloons or tetroons, which have transponders and are tracked over long distances by radar (Peterson 1966; Hoecker 1977; Pack et al. 1978; Reisinger and Mueller 1983). Smaller constant-level balloons are sometimes released as part of large-scale educational programs (Stocker et al. 1990). Trajectory evaluations also make use of long-range tracer observations by aircraft and surface monitoring networks [e.g., Cross-Appalachian Tracer Experiment (CAPTEX)—Ferber et al. (1986); and Across–North America Tracer Experiment (ANATEX)—Air Resources Laboratory (1989) and Draxler et al. (1991)]. In addition, the unfortunate Chernobyl accident left a trail of radioactive isotope observations across Europe that were subsequently used for trajectory evaluations (Klug et al. 1992). However, trajectories estimated from tracer gases are inevitably inferior to trajectories estimated from balloons since it is often difficult to locate the tracer plume centerline.

Many trajectory studies have also been concerned with the effects on the predicted trajectories of differences in space and time resolution of the observed or predicted three-dimensional wind fields (e.g., Rolph and Draxler 1990; Kahl and Samson 1986, 1988). Other research has involved comparisons of trajectories computed by two or more different models, where the definition of model is broadened to include diagnosed or interpolated observed wind fields (Artz et al. 1985; Maryon and Heasman 1988). These studies lead to the conclusion that use of different models and/or input observations can lead to typical 10°–20° differences in trajectory directions, and 100–200 km differences in position of trajectory end points after one day of travel. However, these model-to-model comparisons are inevitably marred by the fact that it is impossible to know which model is "best."

Haagenson et al. (1987, 1990) have compared calculated trajectories with observed trajectories based on tracer gas releases during CAPTEX and ANATEX. The center of the tracer cloud could be identified and followed for about one day during CAPTEX and about three days during ANATEX. The MM4 prognostic meteorological model was used to estimate the wind fields for calculating the trajectories. Figure 6.1 presents a summary of the root-mean-square (rms) separation between observed and calculated trajectories as a function of travel time. The rms separations are about 200, 300, and 400 km after one, two, and three days of travel, respectively. Haagenson et al. (1987, 1990) find that best agreement is obtained when the trajectories are calculated on isentropic surfaces. The rms errors

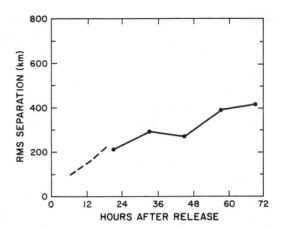

FIG. 6.1. The rms separation (km) between estimated tracer and MM4 modeled trajectories as a function of hours after release. The dashed and solid lines are for the CAPTEX and ANATEX data, respectively (from Haagenson et al. 1990).

are magnified by about a factor of 2 if surface winds rather than winds at the elevation of the tracer cloud are used to calculate the trajectories.

Hoecker (1977) compared calculated trajectories (based on interpolation of observed winds) with observed trajectories of tetroons released from Oklahoma City. Figure 6.2 contains examples of these comparisons for two tetroon flights. The tetroon was recovered at point R and the trajectories calculated using either observed surface winds, observed layer-averaged winds, or geostrophic winds are denoted by curved lines with different symbols. The envelope of the various calculated trajectories includes an angle of about 20°. The calculated trajectories agree better with the tetroon recovery point in the left-hand figure than in the right-hand figure, possibly due to the fact that the calculations in the right-hand figure are complicated by the closeness of the low pressure center. Similar conclusions were reached in the analyses of tetroon trajectories by Reisinger and Mueller (1983) and the analysis of balloon releases by Stocker et al. (1990).

Trajectory analyses similar to those reviewed above have also been reported by Lamb (1981), Walmsley and Mailhot (1981), Warner et al. (1983), Kuo et al. (1985), Gary et al. (1987), Draxler (1987), Brost et al. (1988), Kao and Yamada (1988), and Chock and Kuo (1990). Although space does not permit detailed discussions of these papers, it is clear that there is a consensus among the 20 papers referenced in this section. The following results are supported by most of these papers.

• Root-mean-square angular differences among regional observed and calculated trajectories are about 20°, although the differences can be much greater in the vicinity of fronts and weather systems.

• Root-mean-square differences in distances among the endpoints of regional observed and calculated tra-

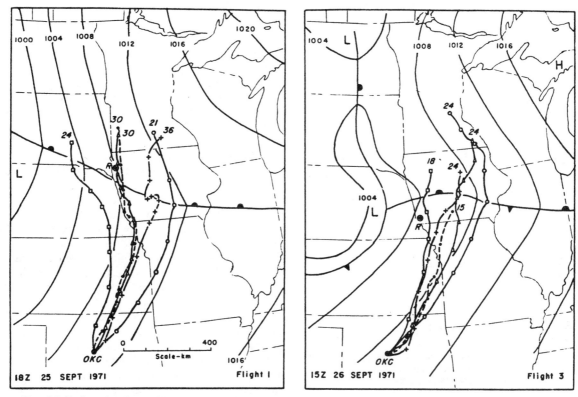

FIG. 6.2. Trajectories observed by two tetroon flights analyzed by Hoecker (1977). The tetroon recovery site is marked by *R*. Time marks along the trajectories are at 3-h intervals and trajectory duration in hours is noted at the end of each trajectory. Fronts and isobars are at about midtrajectory time. The trajectory starting time, which is also the 3-h map time nearest tetroon release time, is shown in the lower-left corner of each map. The varieties of observed wind fields used to calculate trajectories are manual sea level geostrophic vector (circles); layer-average wind by computer (pluses); adjusted surface wind by computer, 2× speed factor [10° veering (triangles), 0° veering (squares)]; and adjusted manual surface wind, 2× speed factor 10° veering (dashes).

jectories are about 100–200 km after one day of travel, and are about 300–400 km after three days of travel. Again, the differences are likely to be larger near fronts and weather systems.

• To calculate trajectories, it is best to use wind data from the atmospheric layer in which the tetroon or tracer gas is traveling. Isentropic trajectories sometimes produce slightly better agreement with observed trajectories than constant-level trajectories; however, caution should be used in calculating isentropic trajectories in frontal regions where the isentropic surfaces are strongly tilted.

• When the space and time resolution of rawinsonde observations of winds are improved, the accuracy of calculated trajectories improves. However, the degree of improvement depends on the complexity of the flow field.

b. Evaluation of mesoscale and regional wind fields

Observed mesoscale wind fields are notoriously variable over time and space. Experimentalists find that if more wind monitors are set up, then more variability is observed. For example, Fig. 6.3 shows the surface wind field observed by about 150 monitors over a portion of Japan covered by a domain of about 300 km × 300 km (Sasaki et al. 1988). This region is characterized by shorelines, mountains, and irregular terrain. Wind direction shifts of 180° and wind speed differences of 5 m s⁻¹ are frequently seen in the data in Fig. 6.3. Even over flat homogeneous terrain, mesoscale turbulence fluctuations can cause variabilities of hourly averaged wind speed 1 or 2 m s⁻¹ over distances of a few kilometers (Lewellen and Sykes 1989; Hanna and Chang 1992).

The wind field evaluation procedures should recognize that certain parts of the domain are inevitably of more interest to the modeler than other parts. For example, if there were an urban area located in the coastal plain in the right-center portion of Fig. 6.3, where the wind patterns are seen to be relatively continuous in space, then the wind field evaluation could be weighted toward those few monitors in that area.

It was mentioned in section 6.2 that statisticians would like independent data to be used for model evaluation. However, with the advent of so-called four-dimensional (three space dimensions plus time) data assimilation (FDDA), newly available data are often

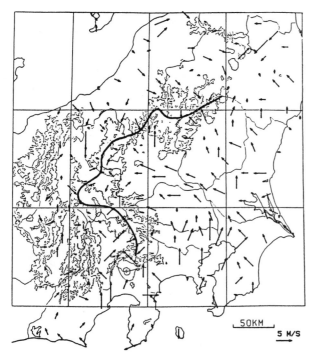

FIG. 6.3. Observed surface wind field over a portion of Japan at 1500 JST 29 July 1983. The thick solid line represents a convergence line between the wind from the Pacific Ocean and that from the Japan Sea (from Sasaki et al. 1988).

used to "nudge" the prognostic meteorological model (Stauffer and Seaman 1987; Anthes et al. 1989). Without nudging or updating with observed winds, after two or three days the predicted wind fields will contain errors due to uncertainties in the initial conditions and boundary conditions, our inability to accurately represent surface characteristics, and model physics and numerics approximations (Anthes et al. 1989). To incorporate the new observations, the modeler applies a nudging "coefficient" that strikes a balance between the desire to force the predicted wind fields to conform to updated observations, and the need to avoid perturbing the model so much that it becomes unbalanced. The nudging term is added to the right-hand side of the equation of motion and has the form $C_1(u_o - u_p)$, where C_1 is the nudging coefficient (usually 0.1 or less) and u_o and u_p are observed and predicted wind speeds.

The data independence issue also arises in evaluations of objective or diagnostic wind field models, where the "modeled" wind fields represent interpolations of the observed wind data. Kessler et al. (1989) and Douglas and Kessler (1991) suggest that if several types of wind observations are available in a region [e.g., National Weather Service (NWS) wind stations, special short-term wind stations, radar sounders, or aircraft measurements], then the model could be evaluated by incorporating all sets of observations except one. Then the simulated winds would be evaluated with the set of observations that was kept out of the modeling ex-

ercise. For example, Fig. 6.4 contains wind fields estimated by Kessler et al. (1989) in the Santa Barbara, California, area at a height of 300 m with a diagnostic model for all observations (rawinsondes) minus a set of aircraft observations [panel (a)], and for all observations including the aircraft observations [panel (c)]. The aircraft observations themselves are the closely grouped vectors plotted in panel (b) of the figure (the other vectors represent rawinsonde observations). It is seen that the magnitudes of the wind speeds (about 5 m s^{-1}) observed by the aircraft in Fig. 6.4b are several times greater than the magnitudes of the wind speeds (about 1 m s^{-1}) in that region produced by the diagnostic model in Fig. 6.4a. As a result, when the aircraft data are included in the diagnostic model, the simulated wind vectors in Fig. 6.4c show a large speed up in the area within about 10–20 km of the aircraft position.

Note that no statistical calculations were carried out with the data in Fig. 6.4, since the differences in the wind fields in panels (a) and (c) are best seen visually (i.e., the look-and-see approach). As another example of this look-and-see approach, panel (a) of Fig. 6.5 contains a plot of the diagnostic wind field for another time period in the same region, and panel (b) contains a plot of the observed dual-Doppler radar wind field in a portion of the region between the coast and the Santa Barbara Islands (Kessler et al. 1989). The radar observations indicate relatively constant northwest flow at speeds of about 5 m s^{-1} over the subregion, whereas the diagnostic wind fields suggest that an eddy exists over that same subregion with both wind speed and direction varying strongly from west to east and north to south. Hundreds of qualitative wind field comparison plots of this type for the Santa Barbara area are contained in the publications by Kessler et al. (1989) and Hanna et al. (1991).

Quantitative evaluations of the predictions of three-dimensional prognostic wind field models with observations and with each other are given by Pielke (1984) and Alpert and Getenio (1988a,b). As mentioned earlier, these evaluations must necessarily give equal weight to all data points. Pielke (1984) presents results of prognostic model runs for sea-breeze scenarios in southern Florida, yielding a typical rms error of about 2 m s^{-1} for near-surface wind speed predictions. The ratio of the rms error to the average standard deviation of the observed predicted wind fields is about 1.0, leading Pielke to conclude that the model was showing some skill. As stated earlier, Keyser and Anthes (1977) suggest that it is desirable that this ratio be less than 1.0 for a good model.

Alpert and Getenio (1988a,b) applied a three-dimensional model and a one-level sigma-coordinate model to the study of surface flow in Israel and compared the predictions of the two models with each other and with a set of detailed observations at a height of 10 m. The flow in the region is complicated due to the influence of a sea-breeze front and the steep mountains

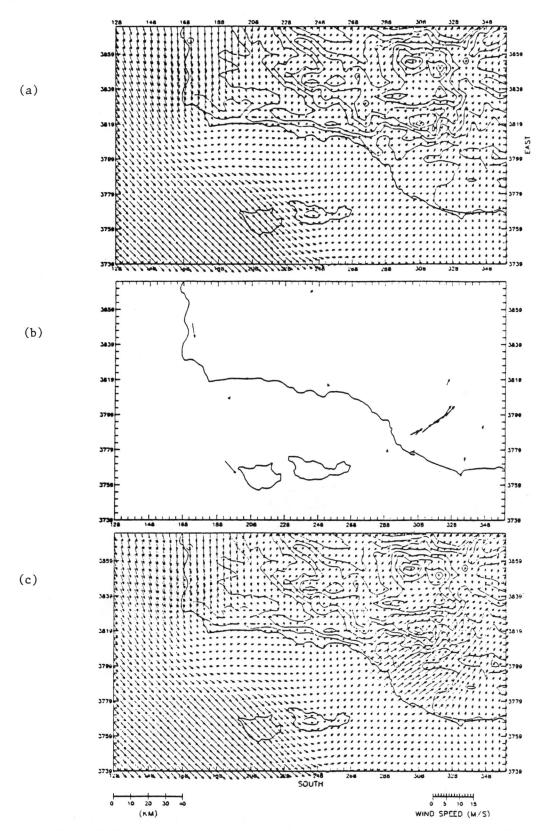

FIG. 6.4. Wind fields calculated by a diagnostic model in the Santa Barbara, California, area at a height of 300 m, (a) incorporating rawinsonde data only and (c) incorporating both rawinsonde and aircraft data. Observed wind vectors for the two types of data are plotted in panel (b), where the aircraft data are seen as the linear series of vectors in the right center (from Kessler et al. 1989).

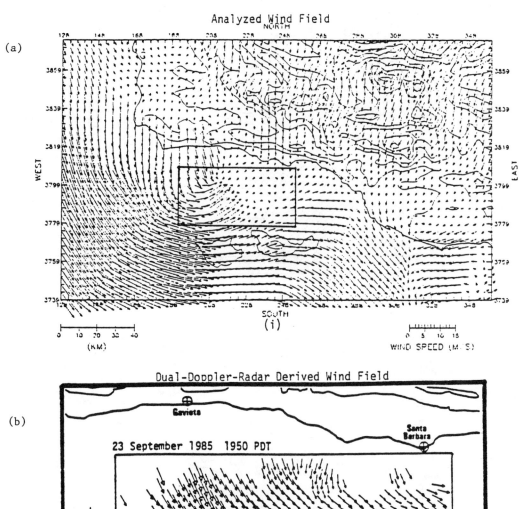

FIG. 6.5. Wind fields estimated by a diagnostic model at 2000 PDT 23 September 1985 over the Santa Barbara, California, region (a), compared with wind fields observed by a dual-Doppler radar in a subregion (b) (from Kessler et al. 1989).

(a)   (b)

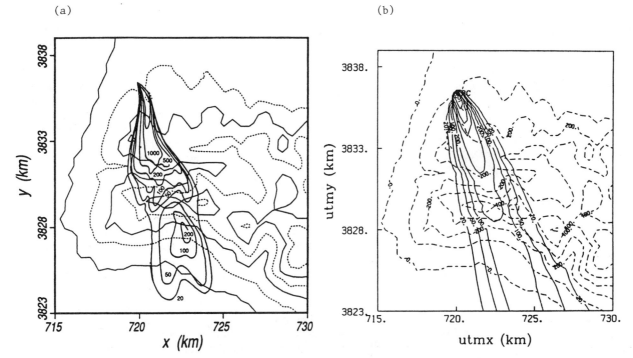

FIG. 6.6. (a) Observed and (b) predicted exposures (s m^{-3}) of inert tracers (fluorescent particles) during Trail MI91 at Vandenberg Air Force Base. The underlying terrain contours are shown at 100-m intervals (from Yamada et al. 1992).

and valleys. The one-dimensional and three-dimensional model predictions were compared with each other qualitatively (i.e., the look-and-see approach) using side-by-side maps of gridded wind vector estimates and observations, leading to the conclusion that the one-dimensional model, with its finer grid resolution, produced results comparable to the three-dimensional model. When the predictions of the one-dimensional model were compared with observations, the rms errors for wind speed and direction were found to be about 1.5 m s^{-1} and 40°, respectively.

It appears that the rms errors in near-surface wind field predictions by mesoscale and regional models are characterized by minimum values of about 1 m s^{-1} for wind speed and 20°–40° for wind direction. Because the magnitudes of these figures are close to the typical natural variability in mesoscale near-surface wind observations over flat terrain (Hanna and Chang 1992), it is unlikely that these rms errors in model predictions can be reduced much further. However, it is important to recognize that the modeled fields are likely to be more reasonable than interpolated fields, in the sense that the space and timescales of the fields are more accurately represented, and phenomena such as sea breezes are better resolved.

c. Evaluation of predictions of mesoscale tracer gas distributions

The predictions of mesoscale and regional meteorological models are frequently used as input to me-

soscale and regional air quality models. A great deal of research has been carried out over the past 10 years in order to evaluate large regional acid deposition and ozone models such as RADM (Dennis et al. 1989) and ADOM (Venkatram et al. 1988), which contain detailed algorithms to simulate the chemical reactions that produce sulfates and ozone. However, for the purposes of the current review, the evaluations carried out with inert tracer gases are more useful (e.g., Draxler 1979 and 1982; Brost et al. 1988; Thuillier 1992; Yamada et al. 1992; Klug et al. 1992). These studies have involved a variety of tracer gases, including fluorescent particles, Kr^{85}, perfluorocarbons, and Cs-137. Although the tracer observations cannot be used to directly verify any specific output parameters (e.g., wind speeds or mixing depths) predicted by the meteorological models, they are useful for implicitly verifying the predictions of variables such as wind velocity and turbulence. For example, Fig. 6.6 contains observations and predictions (by Yamada et al. 1992) of tracer concentrations over a 15 km × 15 km mesoscale domain that includes Vandenberg Air Force Base, California. The mean transport direction, as indicated by the position of the plume centerline, is simulated within ±5%, the plume width (a measure of the turbulence) is fairly accurately simulated, and the magnitudes of the concentrations also match fairly well.

A quantitative measure of model accuracy is the space figure of merit, which is the ratio of the intersection (overlap) of the predicted and observed plume

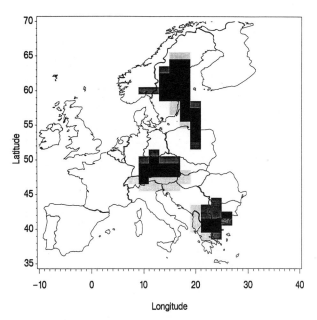

FIG. 6.7. Illustration of calculations of space figure of merit using observations and model predictions of Cs-137 deposition for all days following the Chernobyl accident. The figure of merit (ratio of dark gray area to sum of light gray and medium gray areas) equals 52% for this example, where there are three distinct geographic regions where the precipitation and deposition occurred (from Klug et al. 1992).

areas to the union of the areas. Figure 6.7 contains an example of the predicted and observed areas enclosed by the 5 kBq m^{-2} isopleth for Cs-137 deposition for all days following the Chernobyl accident (from Klug et al. 1992). The space figure of merit is 52% for these data, which indicate three distinct deposition regions (centered over Greece, Austria, and Sweden). The wind fields over Europe were observed to be highly variable in time and in space during this period, with strong vertical wind shears, and rainfall patterns were also variable and often poorly observed.

d. Evaluation of surface fluxes and mixing depths

A few limited evaluations have been published involving meteorological variables other than winds. Mesoscale meteorological models generally include predictions of variables such as turbulent energy, surface sensible heat flux, surface momentum flux, and mixing depths, but these parameters are not routinely printed out and analyzed. However, these variables are required as input to mesoscale and regional air quality models, which are sensitive to their uncertainties. In this section, comparisons are presented of predictions and observations of surface heat and momentum fluxes (Steyn and McKendry 1988; Draxler 1990) and mixed-layer depths (Steyn and McKendry 1988; Ulrickson and Mass 1990).

Draxler (1990) points out that many of the predictions of the NWS Nested Grid Model (NGM) are not

normally saved but could potentially be combined with field observations to produce three-dimensional grids of observations and predictions at 2-h intervals for future use in regional air quality analyses. He made 12-h forecasts with the 91-km grid version of NGM for January, February, and March 1987, which corresponds to the time period when the ANATEX study was conducted. The NGM runs were reinitialized with NWS radiosonde data from the 0000 and 1200 UTC soundings. He compared the model predictions with observations from a tall tower in South Carolina. Because the lowest model level was 175 m, it was necessary to carry out further calculations using Monin–Obukhov similarity theory in order to obtain predictions of surface fluxes.

Draxler's (1990) comparisons of observed and predicted friction velocities (proportional to the square root of the momentum flux) and sensible heat flux are given in Figs. 6.8 and 6.9, respectively. Three-month averages of NGM predictions are given as crosses and observed values from the tower are given as circles. The boxes include plus and minus two standard deviations (i.e., about 95% confidence limits) for the residuals between the predictions and observations, and an asterisk indicates that the means for that hour are significantly different at the 95% confidence level.

It is seen that the 95% confidence limits in Figs. 6.8 and 6.9 are approximately equal to the means and that there are significant differences between the mean predictions and observations on about one-half of the hours. The model has a tendency to overpredict the

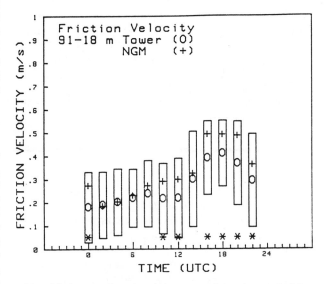

FIG. 6.8. Average diurnal variation over a 3-month period of the friction velocity as estimated from observations of wind speed at the 182-m level of a tower and predictions from the NGM at the first sigma level (175 m), where the friction velocity is derived from Monin–Obukhov similarity theory. The boxes represent two standard deviations of the residual between the measurement and prediction. When the means are significantly different at the 1% level, an asterisk is marked (from Draxler 1990).

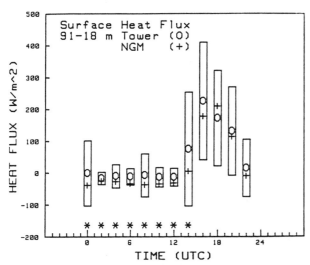

FIG. 6.9. Surface sensible heat flux as estimated from the tower observations of wind speed and temperature difference and the predictions from the NGM, as in Fig. 6.8 (from Draxler 1990).

friction velocities, especially at midday, and appears to overestimate the magnitude of the downward-directed sensible heat fluxes at night. However, it should be remembered that the observations are from a single point, whereas the predictions represent an ensemble mean over the entire 91 km × 91 km grid square.

Figures 6.10 and 6.11 contain predictions of two slightly different versions of the Colorado State University Mesoscale Model (CSUMM). Steyn and McKendry (1988) applied the model to the Vancouver, British Columbia, area and Ulrickson and Mass (1990) applied it to the Los Angeles basin. Predictions are made for a single day. Figure 6.10 contains predictions and observations of sensible heat flux Q_H, ground heat flux Q_G, and latent heat flux Q_E for the Vancouver station, showing good agreement (±10% or 20%) for the sensible heat flux but that the ground heat flux predictions lead the observations by about 3 h while the latent heat flux predictions lag the observations (parameterized) by about 3 h (Steyn and McKendry 1988).

The CSUMM predictions of mixed-layer depth h are compared with observations in Fig. 6.11 [top—Vancouver (from Steyn and McKendry 1988); bottom—Pomona and Cowan Avenue, in the Los Angeles basin (from Ulrickson and Mass 1990]. Observations are from tethersondes in Vancouver and radiosondes in Pomona and Cowan Avenue. The model predictions are clearly about a factor of 2 too high in Vancouver. Predictions are fairly accurate (±10%–30%) at midday at both the inland station (Pomona) and the coastal station (Cowan Avenue) but tend to be in error during morning and evening transition periods. These comparisons are complicated by two prime considerations: 1) it is difficult to observe mixed-layer depth with confidence, and 2) the observations are point measure-

ment, whereas the predictions are ensemble means over a grid square.

Both Steyn and McKendry (1988) and Ulrickson and Mass (1990) apply the model evaluation performance measures suggested by Willmott (1982), which were given earlier in (6.5)–(6.7). They analyze many more variables than we have presented in Figs. 6.10 and 6.11, including surface-layer wind and temperature and vertical profiles of winds and velocity temperature. The index of agreement [d in (6.7)] is calculated to be in the range from about 0.2 to 0.5, suggesting that the model is accounting for about 20%–50% of the variance in the observations.

6.5. Recommendations

Over 50 papers describing evaluations of mesoscale and regional meteorological models are listed in the references, and some of these papers have been dis-

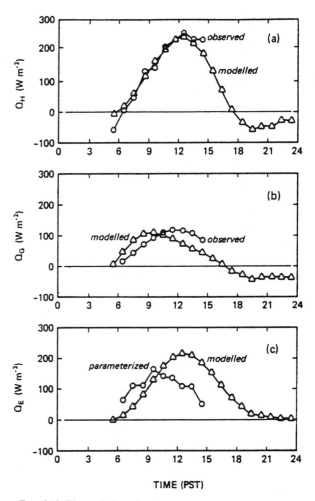

FIG. 6.10. The evolution of modeled and observed (a) turbulent sensible heat flux, (b) subsurface conductive heat flux, and (c) turbulent latent heat flux during one day at a Vancouver, British Columbia, monitoring station. CSUMM is used for the predictions (from Steyn and McKendry 1988).

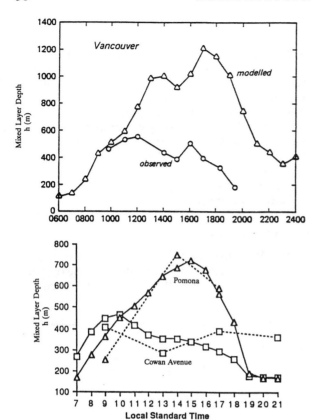

Fig. 6.11. Diurnal variations of modeled and observed mixed-layer depth in Vancouver, British Columbia, (top), and in Los Angeles (bottom, where boxes indicate the Cowan Avenue station and triangles indicate the Pomona station). Observations and predictions are given as dashed and solid lines, respectively, in the bottom figure. The CSUMM is used for the predictions in both figures [from Steyn and McKendry 1988) (top) and Ulrickson and Mass (1990) (bottom)].

cussed in detail in the preceding sections. It is seen that most of the evaluations with field observations have involved qualitative comparisons of plots of wind vectors, trajectories, or tracer gas distributions. In a few cases, quantitative assessments were carried out (e.g., calculations of biases, rms errors, or figures of merit).

Only a few of these papers include more comprehensive analyses of the broad range of meteorological parameters that are important for use as input to regional air quality models. These parameters include turbulent energy, surface energy fluxes, clouds, radiation, mixed-layer depth, and vertical motion. For example, Klug et al. (1992) point out that most of the environmental effects of the Chernobyl accident were due to wet deposition (i.e., rain out) of radionuclides, and the observed distribution of rainfall in space and

time were not as well known as the modelers would have liked. Also, Haagenson et al. (1987, 1990) emphasize that persistent vertical motions (e.g., lifting over a warm front) can strongly influence ground-level tracer patterns. Venkatram et al. (1988) and Dennis et al. (1989) stress that in order to parameterize in-cloud chemical conversion processes, of importance to acid rain modeling, it is obviously necessary to know the location, type, and microphysical characteristics of clouds. Clearly there is much more work to be done in evaluating this full range of meteorological parameters.

Better estimates are needed of the fundamental stochastic uncertainty in the meteorological processes being modeled. It was mentioned earlier that mean winds observed over mesoscale networks have a stochastic or natural variability of about 1 m s^{-1}, even over flat homogeneous terrain. Similar estimates of uncertainties should be made for turbulent energy, radiation, and other parameters of importance. These uncertainties represent the lower limit to model scatter and can also be used to estimate confidence intervals on model predictions.

A fundamental difficulty in evaluation of the predictions of the mesoscale meteorological models is the inherent difference in meaning between the model predictions and the observations. The three-dimensional numerical model predictions represent ensemble averages over grid squares with side dimensions of a few kilometers, while the observations represent single realizations at a point. For strict comparability, many observations should be taken within the grid square for a large number of similar realizations, and the results should be averaged. Methods should be investigated for eliminating as much of this lack of comparability as possible.

Our recommendations for a comprehensive mesoscale meteorological model evaluation program were presented in section 6.3, including both a qualitative peer review of the models' scientific components and a quantitative estimate of performance measures. Perhaps as we gain experience in this area, we can develop so-called model acceptance criteria that can be used to define whether a new model falls within the normal bounds of bias and uncertainty.

Acknowledgments. This research has been supported by the Electric Power Research Institute and the U.S. Army Atmospheric Sciences Laboratory. The author appreciates review comments received from Walter Bach, Roland Draxler, Teizi Henmi, Robert Kessler, R. H. Maryon, and Karl Zeller.

PART II

The Mesoscale Model Comparison Project

Chapter 7

Introduction

ROBERT P. PEARCE

Department of Meteorology, University of Reading, United Kingdom

An extensive body of meteorological data was collected around the Sacramento River valley of northern California during 1985–87 under a cooperative field study conducted by Army Laboratories and the Department of Agriculture Forest Service. Its primary aim was to provide a dataset suitable for validating and further developing mesoscale and terrain-flow models. This study was designated Project WIND (Wind in Non-uniform Domains). Four phases (summer, winter, spring, and fall) of comprehensive meteorological data were collected over complex terrain and variable land use on scales ranging from 200 km × 200 km (mesoalpha) to 5 km × 5 km (mesogamma). Data were collected over a two-week time span and included two full 24-h periods during each phase.

The generation of a fully validated and quality-controlled dataset from the extensive body of raw data was undertaken by Dr. R. Cionco and his group at the Army's Atmospheric Sciences Laboratory (ASL), White Sands, New Mexico. In 1990 it was decided that, in view of the potential value of these data to the several groups in the United States and Europe developing mesoscale models, some already in operational use, an invitation should be issued for participation in a model intercomparison project. ASL would provide data for one 24-h period of summer (phase I) data and one 24-h period of winter (phase II) data, each covering an 80 km × 80 km domain and consisting of data measured from 25 surface sensors and five upper-air sounding stations located within the domain. The surface data would consist of 1-min averages of wind (speed and direction), temperature, and pressure, as well as precipitation information; selected surface stations would also include measurements of incoming solar radiation and temperature. The upper-air soundings, terminating near 350 mb, would be of the standard measurements of wind, temperature, relative humidity, and pressure. Participants would be asked to carry out 24-h forecast integrations with their models for each phase, initializing them with the observed fields at the beginning of each phase and using other data for boundary forcing at their discretion.

The groups expressing interest in participating in the project were sent detailed specifications and formats

of the model outputs required from them, to be submitted to ASL on floppy disks. These outputs were separated into three groups, 3-h horizontal fields, time series at 21 surface stations, and 2-h vertical profiles at each of the five upper-air stations. Surface distributions were required of sensible and latent heat fluxes, horizontal momentum flux, pressure, and precipitation in the last 3 h on a 41 × 41 universal transverse Mercator grid of points at 5-km intervals. At 10 m, on the same grid, 3-h distributions were requested of zonal and meridional wind components, vertical motion, specific humidity, and subgrid turbulent fluxes of momentum, latent heat, and sensible heat. Similar distributions, on a coarser 10-km grid, were requested at 500 and 5000 m.

Time series at the 21 surface stations (hourly values) were requested of temperature, dewpoint, upward and downward components of solar and longwave radiation, sensible and latent heat flux, heat flux into the ground, and precipitation over the last 3 h; zonal and meridional wind components, temperature, dewpoint, and subgrid-scale turbulent momentum fluxes at 2 m and the same, but excluding the turbulent fluxes, at 10 m. Vertical profiles at the five upper-air stations were requested of zonal and meridional wind components, temperature, and dewpoint at 2-h intervals. In view of the large quantity of model output data requested and the limited time available to modelers to carry out the model integrations, the items were given priority indicators to enable groups unable to complete in time to concentrate their efforts on a specified subset of the outputs.

The group at ASL led by Dr. T. Henmi carried out an integration using the phase I data on the model (HOTMAC) being developed there for operational use by the army. They created files of output data in the same format as that requested of the project participants and then developed software to enable the results to be displayed in graphical form. This software, and additional components developed by R. Meyers and his group, were used in the first instance to generate graphs and diagrams in a standard form for a volume prepared at ASL and issued to participants at the Mesoscale Modeling Workshop held in El Paso, Texas, 16–18 June 1992. This preworkshop volume contains a selected subset of results obtained by four participating groups, the same subset for each group. The criteria adopted for this selection were dictated essentially by the meteorological conditions of each of the two phases. Thus phase I was a period during which the diurnal heating cycle played the dominant role and those elements best describing the physics of this cycle were chosen. On the other hand, phase II was dominated by synoptic-scale processes, including a frontal passage, so that the emphasis here was on the flow patterns.

The four contributors were invited to write up their results and conclusions for publication in this volume, and their accounts constitute chapters 9–12. The four models used and implementation procedures are first briefly described in each chapter and then the results for the two phases of the WIND project are presented and discussed. It will be noted that although the four models have much in common they also incorporate some different representations of physical processes; one (FITNAH) is nonhydrostatic, while the others (Tel Aviv, RAMS, and HOTMAC) use a hydrostatic formulation. These accounts are preceded, in the next chapter, by a brief account of the WIND database. Finally, some statistics derived from the results and a short commentary comparing their performance is presented in chapters 13 and 14.

Chapter 8

Overview of the Project WIND Data

RONALD M. CIONCO

United States Army Research Laboratory, White Sands Missile Range, New Mexico

8.1. Introduction

Four phases of Project WIND (Wind in Non-uniform Domains) were successfully conducted in and about the Sacramento River valley of northern California during the period beginning June 1985 and ending October 1987. Project WIND is a cooperative field study conducted by United States Army Laboratories and the United States Department of Agriculture Forest Service. The objective was to collect a comprehensive, meteorological diffusion and dispersion database over complex terrain including land use such as vegetation domains; this would be used to evaluate and improve select meso- and micrometeorological deposition, diffusion, and dispersion models. The Army Laboratories collected all of the meteorological and diffusion data and the Forest Service collected deposition datasets.

8.2. Field study design

The design of Project WIND was based on the multiple-scale data requirements of the Atmospheric Sciences Laboratory's (ASL's) hierarchical system of nested meso- and micrometeorological models (Cionco 1985, 1987). Project WIND was, therefore, designed with nested measurement domains to address scales ranging from mesoscale to boundary layer, surface layer, microscale, and canopy for nested physical domains of 200 km × 200 km × 8 km, 80 km × 80 km × 1 km, and 5 km × 5 km × 0.032 km as shown in Fig. 8.1 (Cionco 1984). More than 54 measurement sites were established within the WIND domain. Only two were major sites collecting data on all scales—one in and about a geometrically neat orchard on flat terrain and the other in and about a complex forest on mountainous terrain. The four phases of WIND were conducted during synoptic meteorological regimes of weak marine incursion, cyclonic activity, shallow convection, and subsidence with the aim of acquiring a database that could be used to test the generality of the model hierarchy. Data was collected over a 2-week time span for selected periods resulting in two full sets of daytime, nighttime, and transition (sunrise and sunset) periods and two full 24-h diurnal periods. Most of the deposition and diffusion trials were run concurrent with these periods or were supported for that specific site and test purpose by the Army Laboratories.

8.3. Measurement sites

The nesting of domains shown in Fig. 8.1 relates to Orlanski's (1975) meteorological scales of mesobeta, mesogamma, and microalpha. Microbeta of the vegetation domains is not discernable in Fig. 8.1 within the microalpha domain. Meteorological measurements were therefore collected on the mesoscale, boundary layer, and microscale domains along with aerial spray/aircraft wake, deposition, diffusion, and dispersion trials in an environment of variable terrain with significant areas of vegetation.

The total meteorological database comprises measurements of wind speed, wind direction, turbulence, temperature, humidity, pressure, solar radiation, soil heat flux, precipitation, the boundary layer structure, and upper-air soundings as well as satellite images. The diffusion and deposition datasets are composed of tracer and smoke concentrations and dosages, aerosol deposition population and drop size, photodocumentation of smoke plume behavior, and aerosol spray patterns and behavior. Data were also acquired of vegetation characteristics of the two major orchard and forest sites.

a. The mesobeta domain

The coarse characteristics of the 200 km × 200 km WIND mesobeta domain are the eastern slopes of the Coastal Range, the broad, flat Sacramento River valley, and the western slopes of the Sierra Nevada. A variety of land usage is present in this domain. The higher elevations are mostly forested; parts of the valley are cultivated (many orchards and rice fields) and parts uncultivated along with extensive grazing areas, small towns (population less than 20 000), and farm communities. Two types of data and sites are required for the mesobeta scale: surface based and upper air. The surface stations are primarily those of USA ASL (45 stations) and some additional stations, CARB, CIMIS,

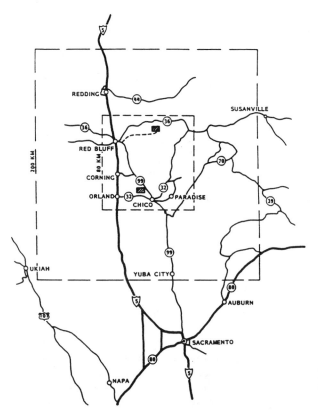

FIG. 8.1. Project WIND region of study.

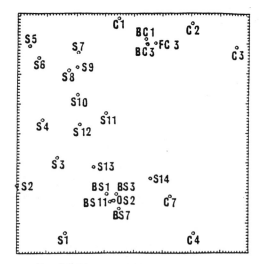

FIG. 8.2. Mesogamma data collection area.

NOWCASTING, and ETAC, with formats and sampling rates for their purposes. The upper-air data were collected by ASL every 2 h during the specified data periods at five sites located totally within the mesogamma domain. Upper-air flights and soundings were terminated at 25 000 ft (about 8 km).

b. The mesogamma domain

The finer topographic characteristics belong to the mesogamma domain (80 km × 80 km), associated with terrain that is generally flat with cultivated fields, orchards, and pastures on the western side near the river, progressing through rolling hills of shrubbery to the forested areas of the western foothills and slopes of the Sierra Nevada. Figure 8.2 locates some of the 54 measuring sites that include surface-based microscale surface-layer and boundary layer instrumentation. Excluding the microalpha stations, 14 automatic surface weather stations (SAMS, designated S1–S14) were located along the river valley and in the foothills, whereas 5 additional surface stations (CAMS, designated C1–C4 and C7) were located widely spread throughout the Sierra Nevada slopes; all used sampling rates of one per second for 1-min averages 24 h daily. Three tethersonde units and one pibal unit were located at four of the five upper-air sites and operated concurrently with the upper-air ascents for elevations of up to 300 m

(tethersondes) and up to 6000 ft (2 km) for pibal flights. Four acoustic sounders were also collocated with the upper-air crews for the same schedule. During phase II, a tethersonde and pibal unit were replaced with two Doppler sodar units. Phases III and IV Doppler sodar units were used at each of the four sites.

c. The microalpha domain

The high-resolution characteristics of WIND are those of two microalpha domains: one in the flat valley and the other in the complex mountain slopes—each collocated with vegetation microbeta domains. Microalpha orchard (5 km × 5 km) is located just northwest of Chico, California, on flat terrain more than half of which is covered with almond orchards and cultivated fields. Figure 8.3 presents the locations of 16 automatic surface weather stations as well as the

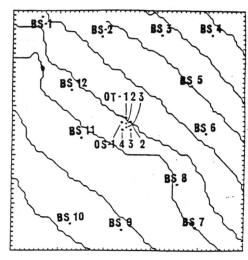

FIG. 8.3. Microalpha orchard domain.

micrometeorological masts for the microbeta scale. Stations BS1–BS12 are located in open terrain of the surface layer, whereas OS1–OS4 are elevated above the almond orchard in the ambient surface layer. All sensors are placed above the ground or canopy surfaces. All were sampled once per second continuously each day during the 15-day period.

Microalpha forest (5 km × 5 km) is some 20 miles east of Red Bluff, California, near Paynes Creek, on forested mountainous slopes with occasional clearings. Nine surface weather stations are located in clearings for surface-layer measurements, elevated above the coniferous forest into the ambient surface layer. All sensors were placed 10 m above ground and some 8 m above the forest surface and sampled once per second for 1-min averages. A third high-resolution study was established during phase IV some 15 miles east of Red Bluff, California, at the Meadowbrook site. Diffusion trials known as AMADEUS (Cionco 1989) were conducted over complex terrain in the ambient surface layer without significant canopy elements. Ten trials were executed.

d. The microbeta domain

Land-use and vegetation characteristics of WIND are those of two microbeta domains collocated in the two microalpha domains described above. The orchard is composed of mature 8-m trees planted on an 8-m-square grid with an adjacent (upwind) extensive, cut grain field of uniform fetch. The three micrometeorological towers are identified as OT1, OT2, and OT3 in Fig. 8.3 of the microalpha scale. Towers of identical sensors and levels are situated outside of the canopy (to provide a reference profile), at the leading and trailing edges inside the canopy and deep into the canopy. Each tower had eight levels of wind components u, v, and w, temperature, and relative humidity; in addition there were heat flux sensors at the bases of OT1 and OT3, net radiation and pressure recorders at OT1, and solar radiation recorders above and below the crown at OT3. The micrometeorological sensors were sampled at one per second during the specific study periods. Wind and temperature were also sampled at higher rates during phases I, III, and IV both in and outside of the orchard for turbulence structure, behavior, and characterization. Vector vane, hot film, and sonic anemometry were used near each of the towers measuring mean values.

The forest, east of Red Bluff, is composed mostly of Ponderosa pines that average 24 m in height together with some Douglas fir and hardwoods on terrain sloping downward from east to west. Three micrometeorological towers (FT1, FT2, FT3) are located outside

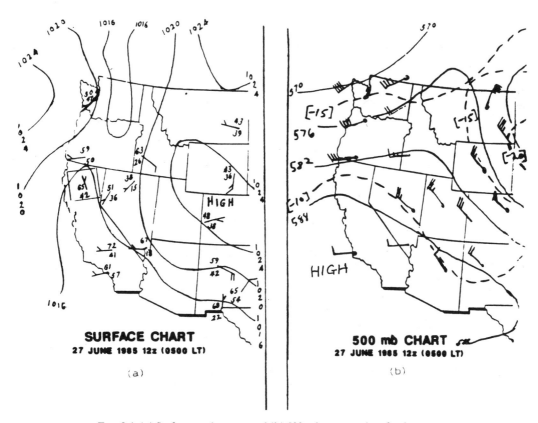

FIG. 8.4. (a) Surface weather map and (b) 500-mb contour chart for the western United States at 0500 LST 27 June 1985.

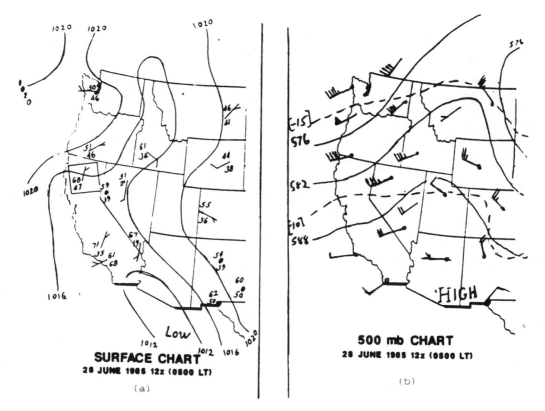

FIG. 8.5. (a) Surface weather map and (b) 500-mb contour chart for the western
United States at 0500 LST 28 June 1985.

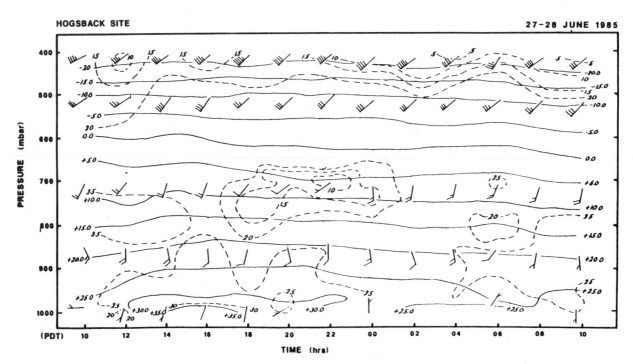

FIG. 8.6. Time section of upper winds, temperature (°C, full lines), and relative humidity (%, dashed lines)
recorded at Hogsback site, 27–28 June 1985.

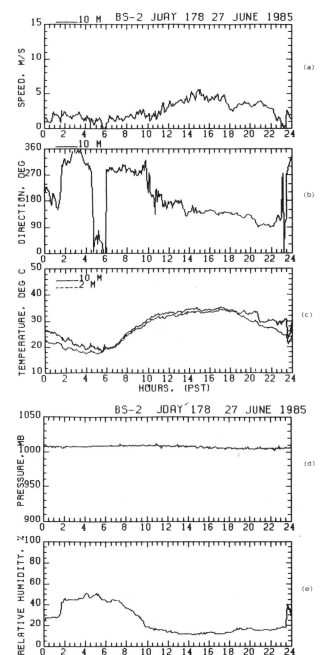

FIG. 8.7. Examples of time series of (a) wind speed (m s^{-1}), (b) wind direction (degrees from north), (c) temperature (°C, 2 and 10 m), (d) surface pressure (mb), and (e) relative humidity (%) recorded at a SAMS site (BS2, 27 June 1985).

of the forest, and others at the leading and trailing edges inside the forest and deep into the forest. Each tower has eight levels (six within the canopy and two above) up to 32 m measuring u, v, w, temperature, and relative humidity as well as heat flux sensors near the bases of FT1 and FT3, pressure at FT1, and solar radiation above and below the tree crown at FT3. The micro-

meteorological sensors were sampled at one per second during the specified study periods. Winds were also sampled at a higher rate during phase IV, both in and outside of the forest, for turbulence behavior and characterization. Hot film and sonic anemometry were collocated on both sides of the canopy edge near FT2 for further turbulence studies through the canopy edge.

Further details of the microscale components of Project WIND are described by Cionco in other papers (Cionco 1984, 1985, 1987, 1989).

8.4. Field study execution

Project WIND is designed to characterize a variety of meteorological conditions, including the differences between day and night, for full model evaluation. It consists of four phases: a weak marine incursion from the bay area (phase I), and periods of cyclonic activity (phase II), shallow convection (phase III), and subsidence (phase IV). Each phase consisted of two daytime (1000–1600 LT), two nighttime (2000–0400 LT), two transition (sunrise and sunset), and two 24-h diurnal periods. These are listed in Table 8.1. The data collection followed the schedule to the minute for each phase, that is, it started on time and finished on time without a miss. A high data recovery rate (93%) resulted in a meteorological database of more than two billion points and on-site quality control ensures that this database is of reasonably high quality.

Four turbulence studies were successfully conducted alongside the basic mean quantity study during phases I, III, and IV in both the orchard and the forest. These databases were obtained for new studies of turbulence behavior at and through the canopy edge. The database was also used for studies of turbulence structure in the vertical to permit comparison of the crown turbulence with that of the trunk space and that of the canopy layer with that of the ambient layer.

TABLE 8.1. Project WIND schedules.

Project WIND field study events:

Jun–Jul 1985	marine incursion	summer
Jan–Feb 1986	cyclonic activity	winter
Apr–May 1986	convection	spring
Sep–Oct 1987	subsidence	fall

Data collection schedule

per event day	Data period (local time)
1	day: 1000–1600
2/3	night: 2200–0400
4/5	diurnal: 1000–1000
6/7	no collection
8	sunset: 1600–2200
9	day: 1000–1600
10/11	diurnal: 1000–1000
12	no collection
13/14	night: 2200–0400
15	sunrise: 0400–1000

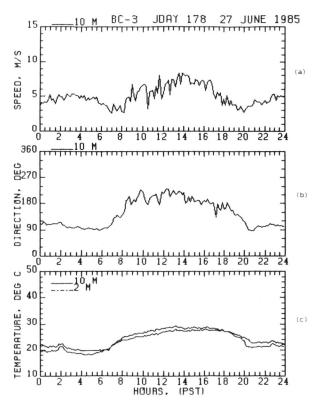

FIG. 8.8. Examples of time series of (a) wind speed (m s^{-1}), (b) wind direction (degrees from north), and (c) temperature (°C, 2 and 10 m) recorded at a CAMS site (BC3, 27 June 1985).

Two field studies of combined aerial spray, aircraft wake, and deposition were also conducted during phases I and III within the two major data collection sites. The datasets—obtained using the same test procedures, equipment, and personel—will allow comparisons to be made of the spraying characteristics over simple uniform and complex nonuniform domains.

Two complete and different diffusion trials were also completed yielding a comprehensive meteorological and diffusion database for a high-resolution domain. These short-period dispersion studies, of diffusion through the forest interface and over complex terrain, will be even more valuable with the addition of the other components of the WIND database.

8.5. Selected datasets

Two 24-h datasets were extracted from the quality-controlled Project WIND database for use in the Mesoscale Model Comparison Project. Datasets representative of a summer day [Julian day (JD) 178–179, 27–28 June 1985] and a winter day (JD 32–33, 1–2 February 1986) were selected and prepared by Army meteorologists. Included in each dataset are 1) upper-air soundings to 25 000 ft for five sites, 2) time histories from 21 surface stations, 3) boundary layer profiles, 4) digitized terrain elevation, 5) site location coordi-

nates, and 6) error flag comments to aid the user. The upper-air soundings were made every 2 h during the 24-h periods starting at 0900 PST on JD178 and 1000 PST on JD32. The surface stations (SAMS and CAMS) operated continuously.

a. Phase I summer dataset, JD 178–179, 27–28 June 1985

Figures 8.4 and 8.5 present the surface and 500-mb charts 0400 PST (0500 LT) for 27 and 28 June 1985. The study area is clearly under the influence of a weak synoptic regime of high pressure resulting in hot dry weather with mostly clear skies and light to variable winds.

For JD 178, the dominant weather feature at the surface (Fig. 8.4a) is a strengthened cell of high pressure centered over northern Colorado moving eastward toward the central plains. High surface pressure is located off the Washington and Oregon coasts. A thermal low extends northward over California. A few scattered high clouds were reported. Temperatures and dewpoints indicate a maritime intrusion through the study domain.

At 500 mb a ridge of high pressure has built up over the northern Rocky Mountain states with moderate zonal onshore flow over the study area. Winds have backed slightly over the previous 24-h period, but there has been little change in upper-air temperatures.

For JD 179, the surface high pressure cell has transferred further eastward (Fig. 8.5a) with a low pressure region positioned off the Washington coast. Temperature and dewpoint over the Project WIND domain are higher than on the previous day, but the winds and cloud cover were reported to be the same.

At 500 mb, the ridge of high pressure persisted over the Rocky Mountain states, but there was an increase in both wind speed and contour height gradient over the Project WIND domain. Temperatures aloft remained virtually unchanged.

Analyses of wind, temperature, and relative humidity for each of the upper-air stations were prepared in the form of pressure–time sections by Project WIND meteorologists. An example, for the Hogsback site, is shown in Fig. 8.6. The 2-h frequency of ascents exhibits variability suggesting the presence of distinctive mesoscale structures even in this virtually static synoptic-scale situation.

The 15 automated surface stations (SAMS 1–15 and BS2) located in the valley and along the foothills collected wind speed, wind direction, temperature, and relative humidity continuously. Figure 8.7 shows example time series at 10 m and, for temperature, also at 2 m. Stations S1–S4 and S12–S14 have the above measurements only at 2 m, but also measure incoming solar radiation.

Six independent surface stations (CAMS C1–C4, C7, and BC3) are located on the western slopes of the Sierra Nevada. Not being part of the automated system, only

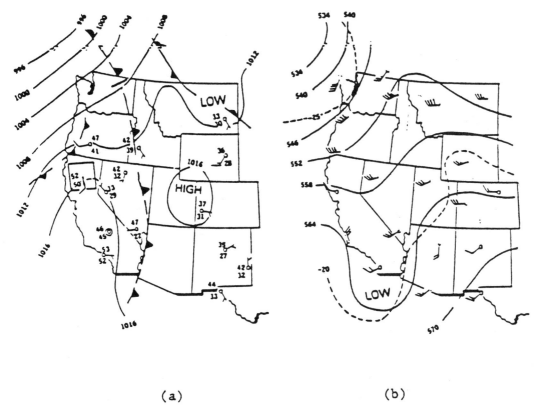

(a)   (b)

FIG. 8.9. (a) Surface weather map and (b) 500-mb contour chart for the western
United States at 0400 LST 1 February 1986.

wind speed, wind direction and temperature at 10 m, with a second temperature at 2 m, are recorded and archived. Figure 8.8 provides an example of the CAMS surface data.

Error flag and quality check comments are also included in the dataset to enable the user to sort through unusual or questionable data points.

b. Phase II winter dataset, JD 32–33, 1–2 February 1986

Figures 8.9 and 8.10 present the surface and 500-mb charts for 0400 PST (LT) for 1 and 2 February 1986. The study area is clearly under the influence of a synoptic event—the approach from the west of a depression with its associated cold, wet weather. The period started with partly cloudy skies, increasing to heavy overcast with the onset of rain and strong gusty winds during the night associated with a frontal passage. This was followed by a return to partly cloudy skies and moderation of the winds. The heavy rainfall and strong (40 kt) winds during the frontal passage prevented several upper-air ascents (a total of 15 between all 5 stations) from being made during 2 February.

On 1 February (Fig. 8.9a) an anticyclone is centered over northeastern Utah. A stationary front extends from northwestern Montana to Oklahoma and a dis-

sipating cold front is positioned from north Washington to western Arizona. Another cold front extends from northwest Washington to the northwest corner of California. The surface wind over the project area is southerly and skies were reported to be completely overcast.

At 500 mb there is a deep trough extending from the Gulf of Alaska to southern California with a ridge centered over the continental divide (Fig. 8.9b). Winds over the project area (which increased in speed over the past 24 h) are westerly.

On 2 February (Fig. 8.10a), the anticyclone persists over Utah and a stationary front extends over northeast Montana. Along the Pacific coast two fronts dominate northern California. One is a dissipating cold front extending from central Oregon through western Nevada; the other is an occluded front running through the Oregon coast into the project area. Strong gusty winds, low overcast clouds, rain, and fog are recorded over the project area. Surface temperatures show little change from the previous day.

At 500 mb, upper troughs continue to dominate the western states. The trough over southern California on 1 February has moved eastward. Over the project area the winds have increased in speed from the previous day and backed to southwesterly. There is little change in temperature.

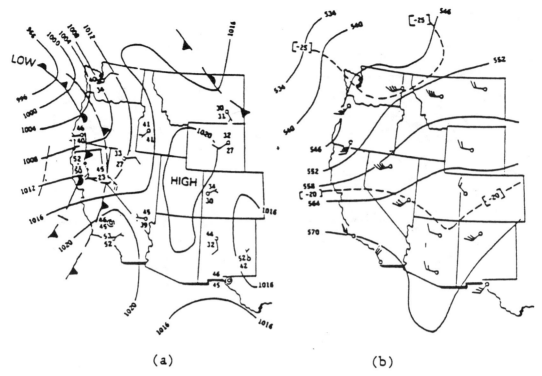

(a)   (b)

FIG. 8.10. (a) Surface weather map and (b) 500-mb contour chart for the western
United States at 0400 LST 2 February 1986.

The forecast (1 February) for the local area for the 24-h period was for a weak front approaching the northwest California coast to maintain mostly cloudy to overcast conditions with light rain showers, followed by strong winds and rain the following day. This forecast underestimated the frontal strength and severity of the weather—that is, precipitation, wind

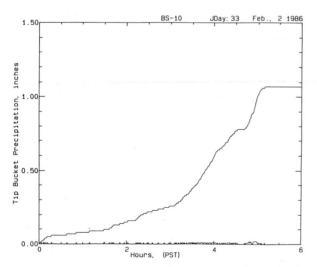

FIG. 8.11. Rainfall recorded at SAMS site BS10, 0000–0600 PST 2 February 1986. The solid curve indicates the accumulation of rainfall (in.); the tick marks along the time axis indicate when rain was falling.

strength, and gustiness, etc. The author's log of events shows that the data collection period started (1000 PST, 1 February) with partly cloudy skies, temperatures in the middle 50s, and light to moderate southeasterly surface winds. By afternoon the low clouds gave way to heavy middle clouds and at 1800 PST scattered showers began from overcast skies. By 2200 PST the winds increased with light rain at some project sites and by 2300 PST all manned sites reported light to moderate rain and, at the foothill sites, gusts exceeding 40 kt. Moderate rain and gusts of 30–40 kt were occurring at all sites by 0130 PST. From 0300 to 0400 PST the front was passing through the area accompanied by heavy rain and strong gusts; all stations missed the 0400 upper-air ascent because of the severity of the weather. Breaks in the cloud started to appear after 0500 PST, and by 1000 PST, the end of the data collection period, skies were again partly cloudy with a light to moderate southeast wind and temperature in the middle 50s. The surface winds did not back to the southwest or farther to the northwest at the frontal passage, as is often observed; it did, however, exhibit notable rainfall and wind speed changes confirming its passage. Figure 8.11 shows the rainfall recorded at the SAMS station BS10 from 0000 to 0600 PST, 2 February 1986.

Because the four phases of Project WIND were all designed to follow the same pattern, the schedules for the upper-air soundings and arrangements for continuous collection of SAMS and CAMS surface data were the same in phase II as in phase I. The selected datasets were prepared in the same way, including error flags and quality check comments. Example plots are not, however, included here for phase II.

8.6. Summary

Project WIND was successfully conducted in and about the Sacramento River valley of northern California during the period beginning June 1985 and ending October 1987. For the purpose of the Model Comparison Project, datasets from phase I, June 1985 and phase II, February 1986 were selected for model initialization and evaluation purposes. This overview describes the design and conduct of Project WIND and the two datasets used by the modelers. These were chosen to be representative of a summer day (27–28 June 1985) and a winter day (1–2 February 1986). Included in each dataset are 1) upper-air soundings to 25 000 ft for five sites, 2) time series for 21 surface stations, 3) boundary layer profiles, 4) digitized terrain elevation, 5) site location coordinates, and 6) error flag comments. Documentation and a user guide were also supplied to the modelers along with the datasets, the latter on magnetic media.

Chapter 9

Project WIND Numerical Simulations with FITNAH

GÜNTER GROSS

Department of Meteorology, University of Hannover, Hannover, Germany

9.1. The numerical model FITNAH

a. Model description

The numerical simulation model FITNAH (flow over irregular terrain with natural and anthropogenic heat sources) is used in the present study. It was developed at the Department of Meteorology, Technical University of Darmstadt, for studies of phenomena belonging to the mesoscale γ. Here, the significant horizontal and vertical scales of the flow are of the same order of magnitude; therefore, the hydrostatic assumption is not valid. Since the nonhydrostatic model FITNAH is described in detail by Gross (1990), only the basic characteristics are repeated here. The framework of the model consists of the equation of motion with Boussinesq approximation, the continuity equation in the anelastic form, the first law of thermodynamics, and an equation for the conservation of water vapor. This is a set of six equations for the three velocity components (west–east component u, south–north component v, vertical component w), potential temperature θ, pressure p, and specific humidity s.

Separating these variables into a mean and fluctuating quantity, correlations of fluctuating velocities and of velocity and scalar fluctuations are introduced by averaging the model equations mentioned above. These turbulent fluxes are parameterized by flux-gradient relationships. This first-order closure method adopted here introduces horizontal and vertical eddy exchange coefficients of momentum, heat, and water vapor.

The vertical one for momentum is determined using the Prandtl–Kolmogorov equation, which gives a relation between the turbulent kinetic energy E and the eddy exchange coefficient. The distribution of E in space and time can be determined by solving a transport equation for this quantity. Such an equation can be derived from the Navier–Stokes equation after some mathematical manipulations. The model equations are given in the appendix.

b. Boundary conditions

A nonslip boundary condition at the earth's surface is adopted for the velocity components ($u = v = w = 0$). Surface humidity is calculated depending on soil water content and the degree of saturation of the lowest atmosphere. The temperature at the ground is determined by a surface energy budget, which includes sensible, latent, and soil heat flux as well as shortwave and longwave radiation. This temperature, and a constant value at 1 m below the surface, are the boundary conditions for the heat transfer equation in the soil.

At the upper boundary at height H, the variables are assumed to be unaffected by the disturbances generated by the topography. This is supported by a damping layer below the material surface top of the model. Therefore, horizontal wind components are set equal to the geostrophic ones, with vanishing vertical wind. Potential temperature and specific humidity are constant in space (horizontal), but variable in time and with height, and are set equal to the larger-scale values.

The formulation of a set of lateral boundary conditions in a limited-area model is difficult, especially when the topography varies significantly along the borders. In the present study, vanishing normal derivatives for all variables are chosen with a correction of the horizontal wind components. This correction is necessary, since vanishing zero gradients do not guarantee mass conservation, an essential requirement for such a long time integration. For this reason, the following procedure is applied. In a first step, the velocity components are calculated corresponding to the boundary conditions mentioned above. Then the total inflow and outflow through the side walls of the integration volume are calculated using the new wind values. Usually, these two values are not identical (partly caused by numerical reasons) and must be corrected by multiplying one of them by a factor such that mass conservation is correct. Inflow and outflow across the lateral borders are now equal. For disagreement with the initial values (for a constant geostrophic wind in time) or with a time-varying geostrophic wind, a further correction is made.

c. Numerical aspects and initialization

The equations are transformed into a terrain-following coordinate system and are integrated forward in time with a self-determined and variable time step. Typical values are 90 s during night and 30 s during unstable daytime conditions. Centered-difference approximations are used in space, whereas an upstream spline technique is used for the advection. To control

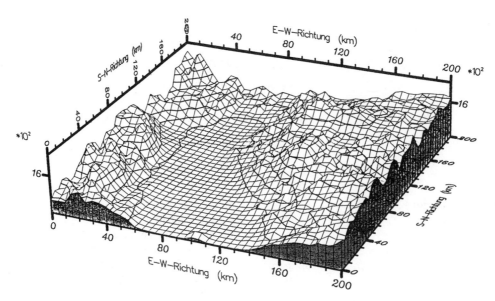

FIG. 9.1. Perspective view of the Project WIND area.

aliasing, a very selective low-pass filter is adopted. The response function of this filter varies with wavelength, such that the short-wave noise is nearly totally eliminated.

The mesoscale pressure is calculated by solving a three-dimensional discrete Poisson equation directly by Gaussian elimination in the vertical and fast Fourier transforms in the horizontal directions.

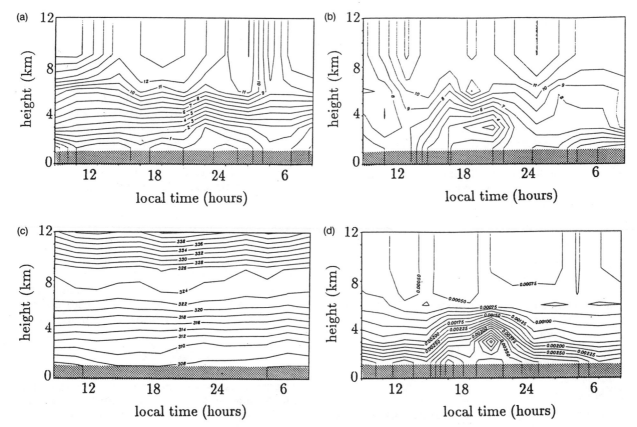

FIG. 9.2. Large-scale input parameters for WIND phase I: (a) west–east component of geostrophic wind, (b) south–north component, (c) potential temperature, and (d) specific humidity.

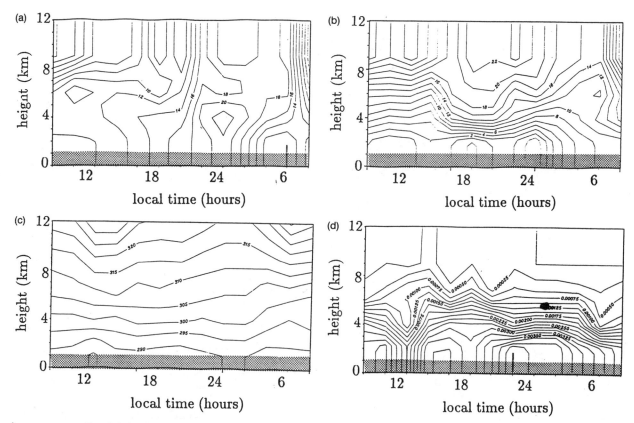

FIG. 9.3. Large-scale input parameters for WIND phase II: (a) west–east component of geostrophic wind, (b) south–north component, (c) potential temperature, and (d) specific humidity.

The model equations are solved on a staggered grid with a constant grid interval of 5000 m in the horizontal directions, while an expanded grid spacing is used in the vertical. The location of these levels are 0, 10, 20, 30, 50, 80, 120, 180, 300, 500, 700, 1000, 1500, 2000, 3000, 4000, 5000, 6000, 7000, 8000, 10 000, and 12 000 m above the lowest point of the terrain. Seven additional levels are below the earth's surface.

The numerical simulation starts with a flat terrain without any horizontal irregularities. The advantage of this procedure is that a consistent set of initial conditions for the variables—namely, the one-dimensional vertical profiles—is known. During the first 15 min of real time the topography is brought in by diastrophy, while temperature at the ground is set equal to the value of the free atmosphere at the same height. For the next 2 h the surface temperature was kept constant and a nearly steady state was obtained. Afterward, the time-dependent simulation was continued but now with a self adjustment of the surface temperature due to the energy budget, whereby the geostrophic wind is kept constant in time for the next hour. After this, the full time-dependent simulation starts with a geostrophic wind, larger-scale temperature, and humidity that are variable both in space (vertical) and time.

d. Input data

The orography is the most important site parameter that determines the distribution of the meteorological variables. The digitized terrain data were provided by the U.S. Army's Atmospheric Sciences Laboratory (ASL) with a 1-km resolution for a 200 km × 200 km area of Project WIND. These values were used to generate the elevation heights on a 5-km grid and used for the simulations in the model comparison. Due to this average, the extremes are smoothed out. The resulting terrain is shown in Fig. 9.1. At all boundaries, an additional two grid points with the same elevation were added (not shown in Fig. 9.1).

The mesoscale region is embedded in a larger-scale weather situation. The mesoscale model (here FIT-NAH) can only simulate the change of this larger-scale situation due to the better resolved local site parameters (e.g., orography). However, the larger-scale situation has to be specified by the space (vertical) and time dependence of the geostrophic wind, large-scale temperature, and humidity. These values were taken from the radiosonde data, where in the lowest 1000 m constant values for the components of the geostrophic wind were assumed. The vertical profile of potential tem-

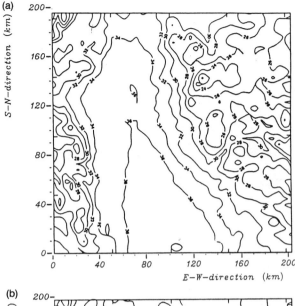

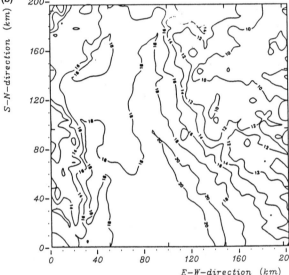

FIG. 9.4. Simulated 10-m temperature at (a) 1500 LST and (b) 0600 LST.

perature is fixed to the value of the standard atmosphere, while relative humidity is kept constant and specific humidity, the variable in the model, is calculated according to this description. The data used in the simulations are given in Fig. 9.2 (phase I) and Fig. 9.3 (phase II).

9.2. Results for WIND phase I

The selected situation for this summer case was that the thermal induced up- and downslope winds were the most dominant feature in the wind field. During daytime, temperatures in the lower part of the valley's atmosphere are relatively uniform and high (up to 36°C). Only the upper parts of the orography are cooler

(Fig. 9.4a). During nighttime, a warmer belt (20°C) is located at the eastern part of the valley, while in the western part temperature is around 2 K cooler (Fig. 9.4b). This feature is in good agreement with the observations. Due to this temperature distribution, a well-developed diurnal wind system has been established. During the day, upslope winds were simulated at all slopes where strong convergences can be found, especially at the mountain peaks (Fig. 9.5a). Inside the valley a divergence line is located. Nearly the opposite pattern is simulated during night (Fig. 9.5b). Strong downslope winds and a convergence inside the valley characterize the wind field.

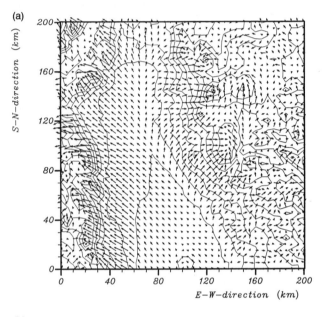

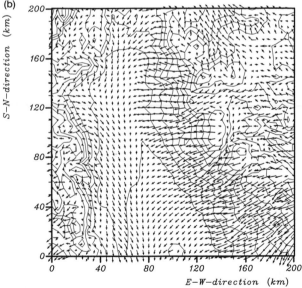

FIG. 9.5. Simulated wind vectors at 10 m AGL at (a) 1800 LST and (b) 0600 LST.

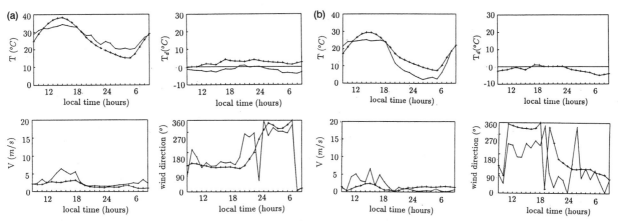

FIG. 9.6. (a) Diurnal variation at 2 m AGL of temperature, dewpoint, wind speed, and wind direction at station S2. Dots indicate observed values and crosses indicate simulated values. (b) Same as (a) but for station S3.

The time variation of wind speed and wind direction, temperature, and dewpoint temperature, all at 2 m above ground, are shown in Fig. 9.6. Two sites were selected for comparison with the observations: one in the valley and one uphill. At the valley station the comparison is reasonable (Fig. 9.6a). Wind speed is generally low and wind direction changes significantly within this 24-h period. The change and the time of wind shift is reproduced very well by FITNAH. The general features of the temperature and dewpoint during the day are simulated reasonably well. The same is true for the uphill station (Fig. 9.6b). While wind speed and wind direction are again well reproduced,

simulated temperatures are in general approximately 5 K higher than those observed. The reason for this might be (a) the badly chosen land use with improper soil characteristics (no information was available) and (b) the elevation on a 5 km × 5 km grid is much lower than the real terrain height. Nevertheless, the overall pattern is reproduced.

The comparison of the observed and simulated vertical profiles of wind and wind direction at the upper-air station 1 are given in Fig. 9.7. The wind speeds at higher levels are too low and the strong fluctuations are not simulated (due to the coarse resolution). However, the general shape is included in the model results. Also, the temperature is lower in the simulations than observed. One can state that the larger-scale values, especially at higher levels, are not accurately prescribed.

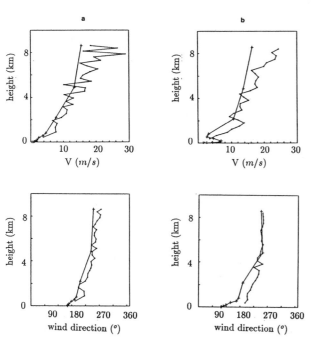

FIG. 9.7. Vertical profiles at upper-air station 1 of wind speed and direction at (a) 1500 LST and (b) 0300 LST. Dots indicate observed values and crosses indicate simulated values.

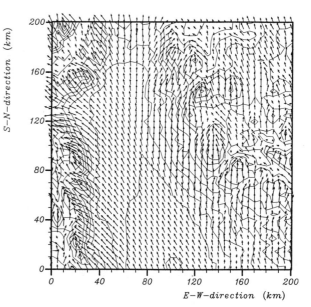

FIG. 9.8. Simulated wind vectors at 10 m AGL at 1800 LST for WIND phase II.

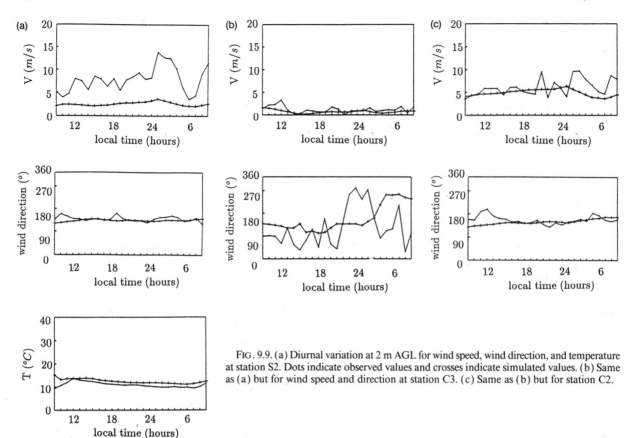

FIG. 9.9. (a) Diurnal variation at 2 m AGL for wind speed, wind direction, and temperature at station S2. Dots indicate observed values and crosses indicate simulated values. (b) Same as (a) but for wind speed and direction at station C3. (c) Same as (b) but for station C2.

9.3. Results for WIND phase II

In contrast to WIND phase I, the winter situation is characterized by a changing synoptic condition with the passage of a frontal trough. Due to this specific weather situation, temperature did not change very much over the 24-h period. Also, the horizontal temperature differences are small and therefore the wind field is dominated by the terrain and, of course, by the changing synoptic condition.

Inside the valley, the airflow is nearly uniform from the south-southeast with modifications at the valley edges (Fig. 9.8). There it seems that weak upslope flows are superimposed on the overall airflow. Comparing the surface observations at site S2 with the simulated values results in a good agreement in temperature and wind direction, while surface wind speed is underpredicted by a factor of 2–3 (Fig. 9.9a). The general tendency to increase until 0100–0300 local time and decrease thereafter is modeled, but the changes in time are much too small. On the one hand, this meteorological variable is well predicted at station C3 (Fig. 9.9b), but the wind direction is poor for the last hours of the integration period. On the other hand, the model results for another uphill station C2, located at the same height above mean sea level but several kilometers to

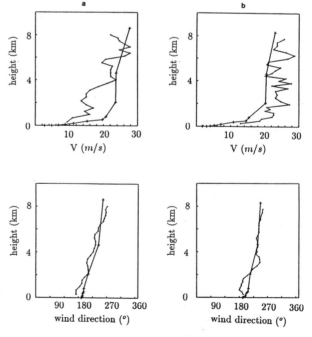

FIG. 9.10. Vertical profiles at upper-air station 1 of wind speed and direction at (a) 1500 LST and (b) 2100 LST. Dots indicate observed values and crosses indicate simulated values.

the northeast of site C3, looks very reasonable (Fig. 9.9c). Wind speed as well as wind direction (at 10 m) are very similar to the observed values.

The vertical profiles of wind speed and direction are given in Fig. 9.10. While wind direction compares well with observations, wind speed shows a more or less similar shape to the measurements, but the values are again too small.

9.4. Conclusions

Two datasets of Project WIND were chosen for a model comparison project. WIND phase I represents a clear summer day, where strong thermally induced wind systems dominate the flow field. In contrast, the winter situation (phase II) is characterized by a weak diurnal temperature variation and the wind field is determined by the orography and the changing synoptic situation. These observations are compared to the results of the nonhydrostatic model FITNAH.

The comparison of the model results with the observed data is very satisfactory and encouraging, especially for phase I. The thermal-induced wind systems are well reproduced by the model in strength as well as their diurnal behavior. The term "well reproduced" implies that the observations are really representative of a larger area since they are compared to a 5 km × 5 km grid averaged value of the model. This point seems to be of special importance for the interpretation of the phase II results. In general, the numerical results for this winter situation are not as good as for the summer case. One reason might be the strong variation of the meteorological variables on short distances caused by the direct surroundings of the stations, but another might be model deficiency. At the present state of the art of FITNAH, it is not possible to handle the passage of a front through the model domain. The larger-scale situation is prescribed, for example, by a geostrophic wind variable with time and (only) with height. While it takes 3–4 h for the real front to pass the WIND region, all grid points in the model domain respond instantaneously to the change of the larger-scale situation.

APPENDIX

The FITNAH Model

a. Equations

First the transformed advection operator and velocity are defined

$$\frac{d}{dt} = \frac{\partial}{\partial t} + u\frac{\partial}{\partial x} + v\frac{\partial}{\partial y} + w^*\frac{\partial}{\partial z^*}, \tag{9A.1}$$

$$w^* = \frac{w}{H-h} - \frac{1-z^*}{H-h}\left(u\frac{\partial h}{\partial x} + v\frac{\partial h}{\partial y}\right). \tag{9.A2}$$

The model velocity equations can be written as

$$\frac{du}{dt} = f(v - v_g) - f^*w - \frac{1}{\rho}\frac{\partial p'}{\partial x} + \frac{z^*-1}{H-h}\frac{\partial h}{\partial x}\frac{1}{\rho}\frac{\partial p'}{\partial z^*} + \frac{\partial}{\partial x}\left(2K_{mx}\frac{\partial u}{\partial x}\right)$$
$$+ \frac{\partial}{\partial y}\left(K_{my}\frac{\partial u}{\partial y}\right) + \frac{1}{(H-h)^2}\frac{\partial}{\partial z^*}\left(K_{mz}\frac{\partial u}{\partial z^*}\right) - n_c c_d buW, \tag{9A.3}$$

$$\frac{dv}{dt} = -f(u - u_g) - \frac{1}{\rho}\frac{\partial p'}{\partial y} + \frac{z^*-1}{H-h}\frac{\partial h}{\partial y}\frac{1}{\rho}\frac{\partial p'}{\partial z^*} + \frac{\partial}{\partial x}\left(K_{mx}\frac{\partial v}{\partial x}\right)$$
$$+ \frac{\partial}{\partial y}\left(2K_{my}\frac{\partial v}{\partial y}\right) + \frac{1}{(H-h)^2}\frac{\partial}{\partial z^*}\left(K_{mz}\frac{\partial v}{\partial z^*}\right) - n_c c_d bvW, \tag{9A.4}$$

$$\frac{dw}{dt} = f^*u - \frac{1}{\rho}\frac{1}{H-h}\frac{\partial p'}{\partial z^*} + g\frac{\theta'}{\bar{\theta}} + \frac{\partial}{\partial x}\left(K_{mx}\frac{\partial w}{\partial x}\right) + \frac{\partial}{\partial y}\left(K_{my}\frac{\partial w}{\partial y}\right)$$
$$+ \frac{1}{(H-h)^2}\frac{\partial}{\partial z^*}\left(2K_{mz}\frac{\partial w}{\partial z^*}\right) - n_c c_d bwW. \tag{9A.5}$$

A continuity equation can be written as

$$\frac{\partial}{\partial x}[\rho(H-h)u] + \frac{\partial}{\partial y}[\rho(H-h)v] + \frac{\partial}{\partial z^*}[\rho(H-h)w^*] = 0. \tag{9A.6}$$

The equation for the potential temperature is written as

$$\frac{d\theta}{dt} = \frac{\partial}{\partial x}\left(K_{hx}\frac{\partial\theta}{\partial x}\right) + \frac{\partial}{\partial y}\left(K_{hy}\frac{\partial\theta}{\partial y}\right) + \frac{1}{(H-h)^2}\frac{\partial}{\partial z^*}\left(K_{hz}\frac{\partial\theta}{\partial z^*}\right) + \frac{1-n_c}{H-h}\frac{1}{c_p\rho}\frac{\partial R_L}{\partial z^*}$$
$$+ \frac{n_c}{H-h}\frac{1}{c_p\rho}\frac{\partial R_N}{\partial z^*} + \frac{n_u}{H-h}\frac{1}{c_p\rho}\frac{\partial R_A}{\partial z^*} - n_c\frac{L_s}{c_p}P_s. \quad (9A.7)$$

The equation for the specific humidity is

$$\frac{ds}{dt} = \frac{\partial}{\partial x}\left(K_{hx}\frac{\partial s}{\partial x}\right) + \frac{\partial}{\partial y}\left(K_{hy}\frac{\partial s}{\partial y}\right) + \frac{1}{(H-h)^2}\frac{\partial}{\partial z^*}\left(K_{hz}\frac{\partial s}{\partial z^*}\right) + n_cP_s. \quad (9A.8)$$

The equation for the turbulent kinetic energy is

$$\frac{dE}{dt} = \frac{K_{mz}}{(H-h)^2}\left[\left(\frac{\partial u}{\partial z^*}\right)^2 + \left(\frac{\partial v}{\partial z^*}\right)^2\right] - \frac{K_{hz}}{H-h}\frac{g}{\bar\theta}\frac{\partial\theta}{\partial z^*} - \frac{E^{1.5}}{a^{-3}l} + \frac{\partial}{\partial x}\left(K_{hx}\frac{\partial E}{\partial x}\right)$$
$$+ \frac{\partial}{\partial y}\left(K_{hy}\frac{\partial E}{\partial y}\right) + \frac{1}{(H-h)^2}\frac{\partial}{\partial z^*}\left(K_{hz}\frac{\partial E}{\partial z^*}\right) + n_cc_dbW^3 \quad (9A.9)$$

and

$$K = 0.2l(E)^{1/2}. \quad (9A.10)$$

---

b. Definitions of symbols

u, v, w^*	wind components
z^*	transformed coordinate $[=(z-h)/(H-h)]$
h	height of orography
H	top of model atmosphere
c_d	effective drag coefficient for plants
b	vertical profile of leaf area density
R_L	net radiation flux without trees
R_N	net radiation flux inside canopy
R_A	anthropogenic heat flux in urban areas
n_c	fraction of a reference area covered by trees
n_u	fraction of a reference area covered with buildings
E	turbulent kinetic energy
K	eddy viscosity
l	mixing length
$(\ )'$	mesoscale deviation

c. Solution technique

The solution is found by using finite differences. The Arakawa C grid is used to locate the variables and perform differencing. The time differences are forward or Adams–Bashforth. The space differencing is centered for all terms except advection, which uses upstream differencing or upstream spline. No numerical smoothing is used in the upstream differencing; however, a low-pass filter is used on the upstream spline calculation.

At the earth's surface a nonslip boundary condition is used for the three velocity components ($u, v, w = 0$). Prandtl theory is used to derive the turbulent kinetic energy relation $E = u_*^2/0.2$ (u_* is the friction velocity). Surface humidity is calculated from soil water content and the degree of saturation in the lower atmosphere.

The radiation computation is performed as follows. The longwave radiation is computed from the relation

$$\frac{1}{H-h}\frac{1}{c_p\rho}\frac{\partial R_L}{\partial z^*} = \frac{1}{H-h}\frac{1}{c_p\rho}\frac{\partial}{\partial z^*}(R_{L\Downarrow} - R_{L\Uparrow}). \quad (9A.11)$$

Here, $R_{L\Uparrow}$ and $R_{L\Downarrow}$ are the downward and upward fluxes at each level. These values are computed from the following equations

$$R_{L\Downarrow} = \int_z^{z_T} \sigma T^4 \frac{\partial\epsilon}{\partial z'}\,dz', \quad (9A.12)$$

$$R_{L\Uparrow} = \sigma T(h)^4[1 - \epsilon(z, h)] + \int_h^z \sigma T^4 \frac{\partial\epsilon}{\partial z'}\,dz'. \quad (9A.13)$$

In these equations, the water vapor atmosphere is assumed to be 15 km high, and empirical formulas are used for emissivities. An empirical formula is used to calculate the heating and cooling of the atmosphere due to shortwave radiation.

Chapter 10

Project WIND Numerical Simulations with the Tel Aviv Model: PSU–NCAR Model Run at Tel Aviv University

P. ALPERT AND M. TSIDULKO

Department of Geophysics and Planetary Sciences, Tel Aviv University, Tel Aviv, Israel

10.1. Description of the model

The model was developed by Anthes, Warner, Ying-Hwa Kuo, and their colleagues at the National Center for Atmospheric Research (NCAR) and The Pennsylvania State University (PSU).

Equations: The version of the PSU–NCAR Mesoscale Model (MM4) that was used in the studies is described along with the full set of equations in Anthes et al. (1987) and Hsie (1987). These are summarized in the appendix. The model is hydrostatic. The closure physics on the subgrid are first order, and detailed boundary parameterization is according to Blackadar's formulation. A grid-scaled diffussivity is also included for stability.

Dimensionality: One-, two-, and three-dimensional versions of this code have been developed.

Grid: Typical grid spacing of this code is 10–80-km resolution, but some runs have been made with resolution down to 5 km. In the vertical, grid spacing is from 30 m near the ground to 1 km higher up. The standard version has 16 layers, but more levels can be used. The horizontal domain is variable and is typically less than 5000 km on a side. The grid is a staggered Arakawa B grid. The coordinate system uses height pressure coordinates in the vertical and either polar stereographic, Lambert conformal, or Mercator map projections.

Analysis methodology: The analysis methodology uses National Meteorological Center (NMC) or European Centre for Medium-Range Weather Forecasts (ECMWF) analyses enhanced by significant and standard-level radiosonde data plus artificial (subjective) soundings generated by satellite data. Here, only Project WIND radiosonde data were used.

Model initialization procedure: Model initialization procedures include normal-mode initialization, boundary nudging, balance equations, 4D data assimulation, and adiabatic and diabatic processes.

Solution technique: The finite-difference method approximately conserves mass, momentum, and total energy. The staggered grid was shown to give signifi-

cantly better results in tropical cyclone simulations; increased accuracy was attributed to better pressure gradient and horizontal divergence calculations. The $p*u$ and $p*v$ variables are calculated at the so-called dot points, and all other variables are calculated at offset cross points. Several options exist for boundary treatment. The usual upper boundary condition is $\dot{\sigma} = 0$. There are five lateral boundary condition options: fixed, time dependent, time dependent in/outflow, sponge, and Davies relaxation.

In the following, only fixed (summer) or time-dependent (winter) options were used.

Surface boundary treatment: The surface boundary treatment enforces $d\sigma/dt = 0$.

Soil treatment: Ground temperature is obtained from a surface energy budget. Over land, the surface energy budget of Blackadar (Zhang and Anthes 1982) is used to calculate the ground temperature through a temperature force/restore slab soil model.

Canopy and land-use treatment: Momentum and energy parameterizations consistent with Blackadar's planetary boundary layer (PBL) are used for a variety of land-use and vegetative combinations.

Special conditions: The model has been run under various conditions as described by Anthes (1990).

Radiation parameterization: Shortwave transmitivity is according to Benjamin (1983), and cooling rate treatment is according to Paltridge and Platt (1981).

Hydrological cycle: Four options exist for the moisture handling: vapor treated as a passive variable; excess water vapor over RH_c removed as precipitation; cumulus parameterization following Kuo (1974) and Anthes (1977), with total convective heating proportional to the moisture convergence in a vertical column (this is the option used here); and an explicit prognostic scheme for water vapor, cloud water, and rainwater (Hsie and Anthes 1984).

Coding practices: Fortran 77, vectorization, modularity, Cray XMP, YMP, CDC 990, IBM 320, and visualization on a workstation. Here, the model was run on the IBM workstation.

SURFACE WIND   27/6/85   17GMT + 21 hrs   06 LST, DAY 179

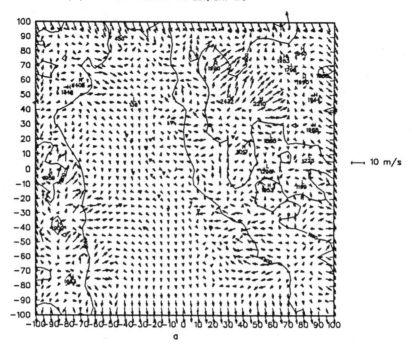

a

SURFACE WIND   27/6/85   17GMT + 6 hrs   15 LST, DAY 178

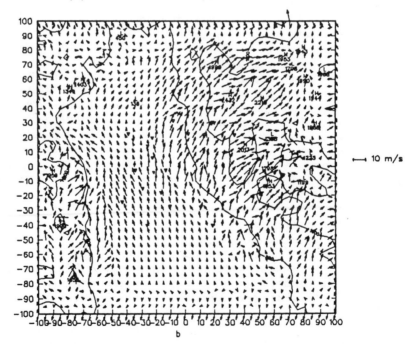

b

FIG. 10.1. Surface wind vectors for (a) 0600 and (b) 1500 PST 27 June 1985 over the full simulation region, in Sacramento Valley. Topographical contours of 800 and 1800 m are shown. Wind scale is indicated on the right.

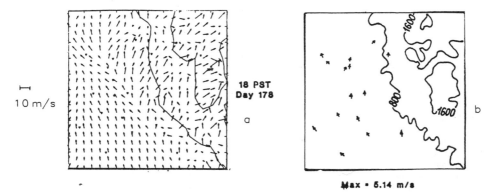

FIG. 10.2. Surface wind vectors from (a) model and (b) observations at 1800 PST over the inner
80 km × 80 km domain. Topographical contours as in Fig. 10.1. Wind scale indicated.

10.2. Setup for WIND simulations

In the following, MM4 version 7 as described by Anthes et al. (1987) was applied, with 16 σ levels at the approximate half σ levels of 0, 5, 15, 28, 55, 230, 450, 920, 1900, 3200, 4900, 6600, 8500, 10 700 and 14 000 m. Top model pressure p_t was 10 mb and a 10-s time step was necessary with the 5-km horizontal interval. Ten computer hours were required for a grid of

41 × 41 × 16 run on a workstation IBM 320 for a 24-h simulation.

For the initial fields the four or five available radiosondes (summer or winter) were interpolated. No use of any surface data was attempted at this stage. The lateral boundary conditions were fixed or updated (in the winter phase) where the time variation followed the 12-h interpolated fields based on the five available radiosondes.

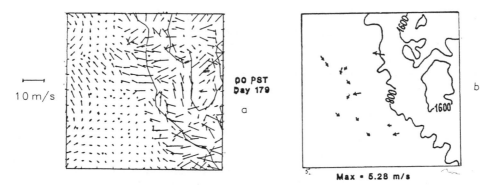

FIG. 10.3. As in Fig. 10.2 except for 0000 PST.

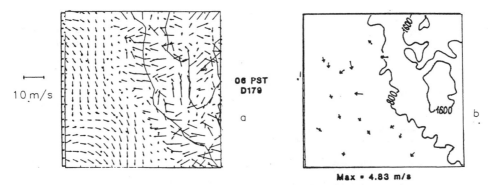

FIG. 10.4. As in Fig. 10.2 except for 0600 PST.

TABLE 10.1. Description of land-use categories and physical parameters for summer
(15 April–15 October) and winter (15 October–15 April).

Land-use integer identification	Land-use description	Albedo (%)		Moisture availability (%)		Emissivity (percent at 9 μm)		Roughness length (cm)		Thermal inertia (cal cm^{-2} K^{-1} s$^{-1/2}$)	
		summer	winter	summer	winter	summer	winter	summer	winter	summer	winter
1	Urban land	18	18	5	10	88	88	50	50	0.03	0.03
2	Agriculture	17	23	30	60	92	92	15	5	0.04	0.04
3	Range-grassland	19	23	15	30	92	92	12	10	0.03	0.04
4	Deciduous forest	16	17	30	60	93	93	50	50	0.04	0.05
5	Coniferous forest	12	12	30	60	95	95	50	50	0.04	0.05
6	Mixed forest and wetland	14	14	35	70	95	95	40	40	0.05	0.06
7	Water	8	8	100	100	98	98	0.0001	0.0001	0.06	0.06
8	Marsh or wetland	14	14	50	75	95	95	20	20	0.06	0.06
9	Desert	25	25	2	5	85	85	10	10	0.02	0.02
10	Tundra	15	70	50	90	92	92	10	10	0.05	0.05
11	Permanent ice	55	70	95	95	95	95	5	5	0.05	0.05
12	Tropical or subtropical forest	12	12	50	50	95	95	50	50	0.05	0.05
13	Savannah	20	20	15	15	92	92	15	15	0.03	0.03

10.3. Summer phase I results

Only four radiosondes were available at the initial time (1000 PST) on the 27 June 1985. The lateral boundary conditions were kept constant.

a. Wind simulations

Figures 10.1a and 10.1b show the surface wind results for 0600 and 1500 PST corresponding to 21 and 6 h of simulation. The total domain is shown and the katabatic (downvalley) flow at night as well as the an-

abatic (upvalley) flow during day are simulated. Comparison with observations for the inner 80 km × 80 km domain is shown in Figs. 10.2a,b, 10.3a,b, and 10.4a,b for 1800, 0000, and 0600 PST, respectively. The model is capable of predicting some small-scale flow features like wind turning from southeast to southwest over the valley as moving from south to north at 1800 PST (Figs. 10.2a,b). Also, a relatively strong easterly flow (~8 m s^{-1}) into the valley at midnight is predicted. Other features are not so well reproduced. For instance, the location of the southeast–northwest convergence line in the valley seems to be predicted slightly farther to the east (~10 km). In the following sections, some of the model sensitivities to land-use variability, subscale topography, and moisture initialization are explored.

b. Land-use variability

Figure 10.5 shows the land-use categories over the model domain as given by the NCAR land-use file. Resolution of the land-use (LU) set is coarser than the model grid (1/6° × 1/4°). Most of the region is characterized by LU category 5—coniferous forest; LU = 4—deciduous forest; LU = 3—grassland or LU = 2—agricultural area. Table 10.1 summarizes the physical parameters for the various categories. Two simulations were performed: one with a constant LU category 3, and the other with the variable LU as in Fig. 10.5. Figures 10.6a,b and 10.7 show the surface latent and sensible heat fluxes (LH and SH) and the Bowen ratio (BR), respectively, where

$$BR = \frac{SH}{LH},$$

for the constant LU run, and Fig. 10.8 shows the BR field for the variable land use. One can notice drastic changes in the BR field over some regions. For instance,

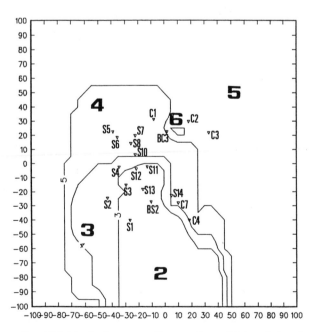

FIG. 10.5. The land-use categories over the model domain based on NCAR classification over a grid of (1/6° × 1/4°). Stations' locations are indicated by triangles. Physical parameters for each category are in Table 10.1.

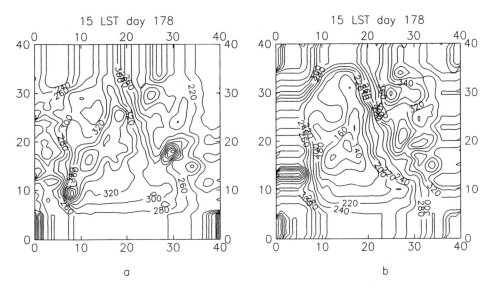

FIG. 10.6. The surface (a) latent and (b) sensible heat fluxes (W m^{-2}) at 1500 PST, for the constant land-use run.

over the southern part of the valley (grid point 20, 4) BR value drops from 0.9 in the constant LU run (Fig. 10.7) to 0.4 with the variable LU. The reason for this is clear; the constant LU category 3 replaces over this location the moister agricultural area with LU = 2, and so the surface moisture availability is too low (15% instead of 30%, Table 10.1). Consequently, the LH is underestimated leading to a near doubling of BR. Obviously, there are other effects like the too low roughness (12 cm instead of 50 cm, Table 10.1) with LU = 4, where the range grassland replaces the deciduous

forest in the constant LU = 3 run. The partition between LH and SH is most important in driving the thermally induced circulations over the valley, and a doubling of the Bowen ratio certainly represents a major change in this partition.

c. Topographical resolution

Figures 10.9a–d and 10.10a–d show the 10-m diurnal variations for stations S10 and S11. Although the two stations are quite close, separated by only about 10 km

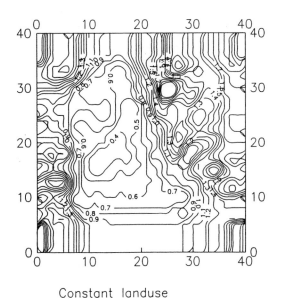

Constant landuse

FIG. 10.7. The Bowen ratio (SH/LH) for the constant land use at 1500 PST. Coordinate numbers indicate grid values.

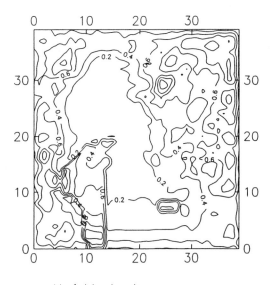

Variable landuse

FIG. 10.8. As in Fig. 10.7 except for the variable land-use run as shown in Fig. 10.5.

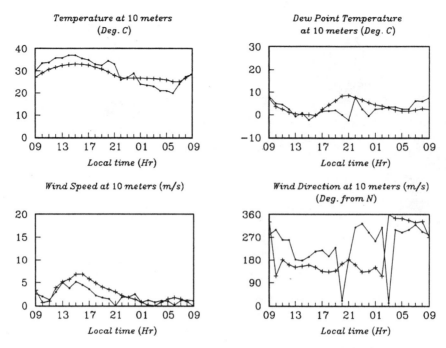

FIG. 10.9. Diurnal variation of (a) temperature, (b) dewpoint temperature, (c) wind speed, and (d) wind direction for both model (crosses) and observations (dots) at station S10. Units are indicated.

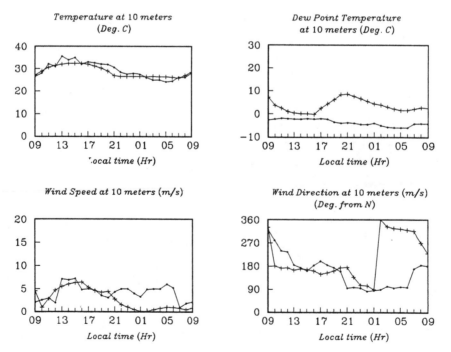

FIG. 10.10. As in Fig. 10.9 except at station S11.

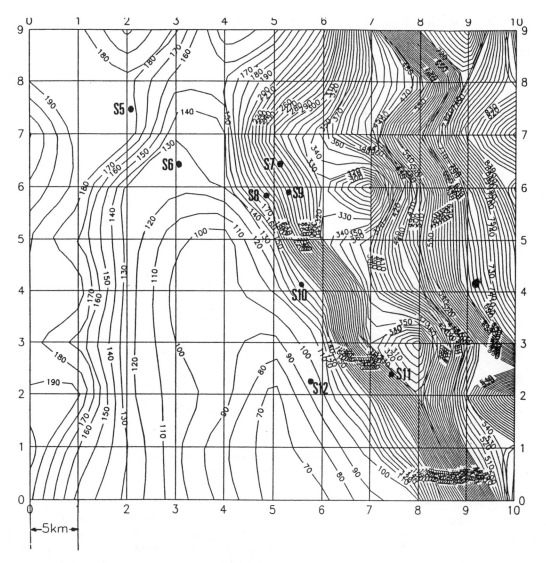

FIG. 10.11. Blowup of the topography at the neighborhood of stations S10, S11. Contour interval is 10 m.

(about two grid intervals), the model verification against observation is very different. This is particularly noticeable through the much drier and the stronger nocturnal wind speeds in S11. The model seems to represent the topography near S10 better. The blowup of the topography (Fig. 10.11) shows that a 5-km resolution will not suffice if differences between the two stations are required in the model simulation.

d. Moisture initialization

The initial surface moisture field at the first time step ($t = 10$ s) along with observed surface relative humidities are shown in Fig. 10.12. It indicates that the surface observed humidity is lower than that deduced from the radiosonde initialization. The lowest-level RH values from the radiosondes were 23%, 23%, 28%, and 41%, whereas the nearby observations show

lower values near 15%. It was suggested that radiosonde RH measurements were less accurate than for the surface stations.

e. Vertical resolution

Figure 10.13 compares the model and the observations at upper-air station 1 for 1500 and 0300 PST on the upper and lower panels, respectively. The fields are (from left to right) wind speed, wind direction, and dry and dewpoint temperatures. As expected, the upper-level dewpoint inversion at approximately 7 km is not predicted, since the model resolution at that altitude is about 1.5–2 km. Another point of interest is the relatively weak upper-level wind intensities at both 0300 and 1500. These deviations are larger at night following 18 h of simulation and are attributed to the increase of the synoptic upper-level wind intensities

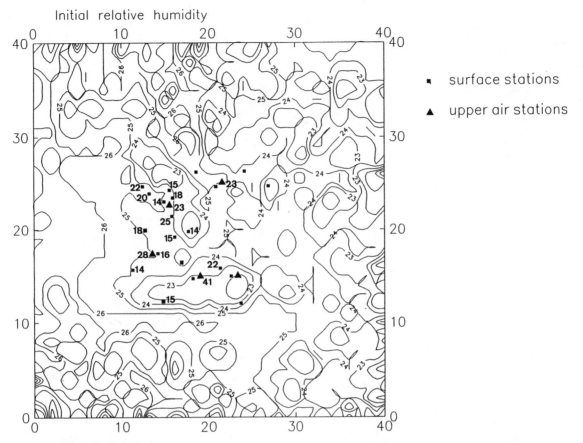

FIG. 10.12. Surface relative humidity field at first time step (10 s) along with the observed values at surface (squares) and upper-air (triangles) stations.

during the day. These were not accounted for in the constant lateral boundary summer run.

10.4. Winter phase II results

All five radiosondes were available for interpolating the initial winds as well as for updating the lateral boundary conditions each time step. For the time interpolation only, the radiosondes at 1000, 2200, and 1000 PST on 1–2 February 1986 were employed, that is, with a 12-h interval. Again, as in the summer, some of the model sensitivities are explored: subscale topographical features, lateral boundary conditions (BC), and vertical wind interpolation.

a. Subscale topography

Figures 10.14a and 10.14b show the model verification for the diurnal temperature variation, at stations S5 and S6, respectively. As indicated earlier, for the similar case in summer (Figs. 10.9 and 10.10) the observations differ widely, whereas the model predictions show similar diurnal variations. Figure 10.15 shows the topographical blowup of the area illustrating the

very different exposure of S5 and S6, although their separating distance is about one grid length only. Obviously, higher horizontal resolution is required in the model in order to capture these subscale variations.

b. Lateral boundary conditions

In this case where a weak frontal passage was observed past midnight, the update of the lateral BC was found necessary, but insufficient, for a successful prediction. For instance, Figs. 10.16a and 10.16b show the model verification for the diurnal wind speed variation at station S4 with constant and variable lateral BC, respectively. The observed double peak wind increase after midnight associated with the frontal passage is in sharp contrast with the model wind weakening when a constant BC is employed (Fig. 10.16a). In the variable BC, however, the model does show some speed increase ($\sim$4 m s^{-1}), but this is still quite far from the observed values of about 7 m s^{-1}. Similarly, large deviation is noticed in the model wind direction at S3 (Fig. 10.17a) with a constant BC. This is considerably improved when the variable BC was employed (Fig. 10.17b).

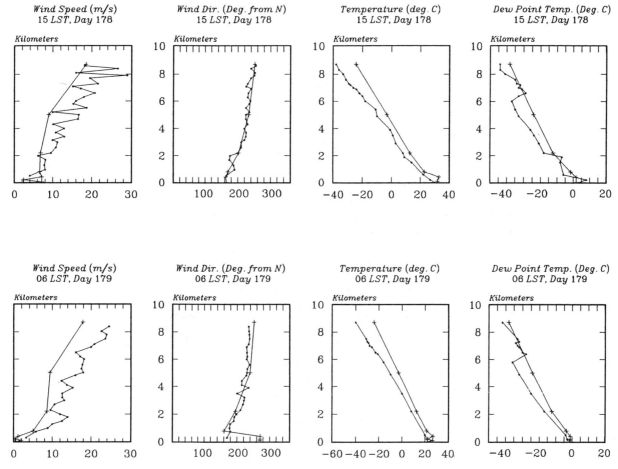

FIG. 10.13. Vertical profiles of (from left to right) wind speed, wind direction, temperature, and dewpoint temperature for both model (crosses) and radiosonde observations (dots). Upper and lower panels are for 1500 and 0300 PST, respectively. Units are indicated.

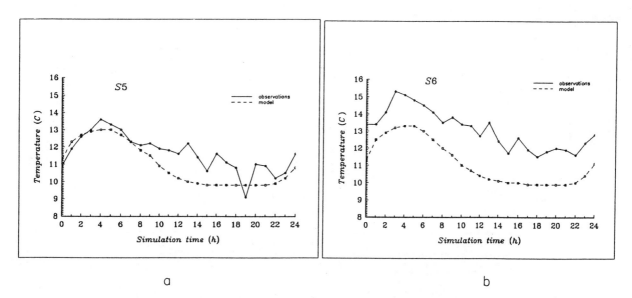

a   b

FIG. 10.14. Verification of diurnal temperature variations at stations (a) S5 and (b) S6.
Observations (solid) and model (dashed) results are shown.

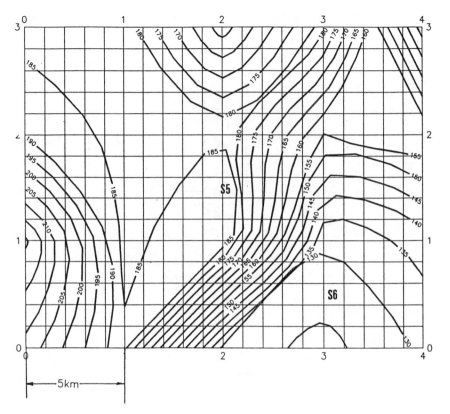

FIG. 10.15. Blowup of the topography in the neighborhood of stations S5 and S6.
Contour interval is 5 m.

c. Vertical wind interpolation

When verifying model predictions we noticed that the way that vertical interpolation from the σ or p levels is performed is quite important, particularly near the surface. Figures 10.18a and 10.18b show the 2-m wind speed verification at station S4 with linear and logarithmic interpolations from 5 m, respectively. Indeed, the linear interpolation weakens the wind speed too much (Fig. 10.18a), whereas the logarithmic interpolation does well. As already discussed earlier (Fig. 10.16), the further deviations later into the simulation

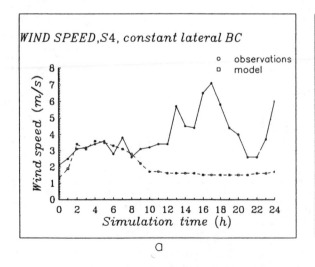

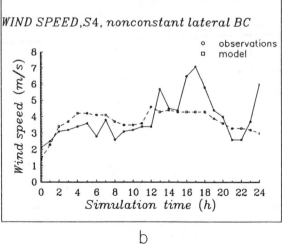

a   b

FIG. 10.16. Model verification for diurnal wind speed variation at station S4 with (a) constant and (b) variable lateral boundary conditions. Model (dashed) and observation (full) curves are drawn.

ALPERT AND TSIDULKO   91

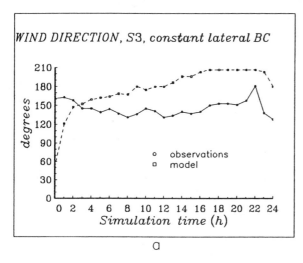

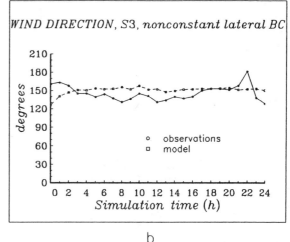

FIG. 10.17. As in Fig. 10.16, but for the wind direction at station S3.

are due to the constant BC that was employed in this simulation.

d. Spatial moisture variability

One of our findings is the large variability in the surface moisture verification. Figures 10.19a–h show the relative humidity verification at stations S1, S2, S3, S4, S5, S6, S14, and BS2. In the first four, agreement of prediction with observations is good, but this is not so for the four stations over the right panel. This is not surprising in light of the large spatial variability of surface moisture. Also, the geographical distribution for these two station groups is different (Fig. 10.5). The better predicted group is at the southwest part of the valley.

10.5. Separating the effects of terrain and surface fluxes

Stein and Alpert (1993) have recently suggested a method that allows the separation of the effects of various processes and their interactions in the model output. They showed that for n processes 2^n simulations are required for a complete separation. The method was applied to the 1500 PST wind vectors of the summer simulation where two processes were chosen: the surface fluxes and the topography. Each vector in the control run was decomposed into the four following vectors: terrain ($\mathbf{T}$), fluxes ($\mathbf{F}$), terrain–fluxes interaction ($\mathbf{TF}$), and the residual vector mainly associated with the large-scale effect (designated by $\mathbf{LS}$).

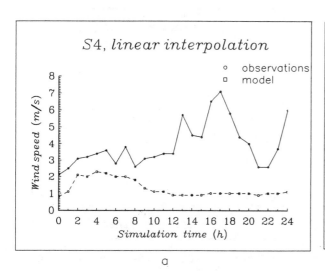

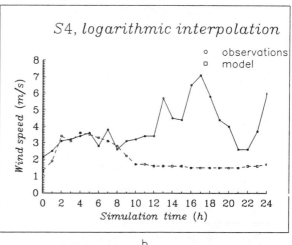

FIG. 10.18. Model verification for the 2-m diurnal wind speed at station S4 with (a) linear and (b) logarithmic interpolations from the 5-m model prediction. Model (dashed) and observation (full) curves are drawn.

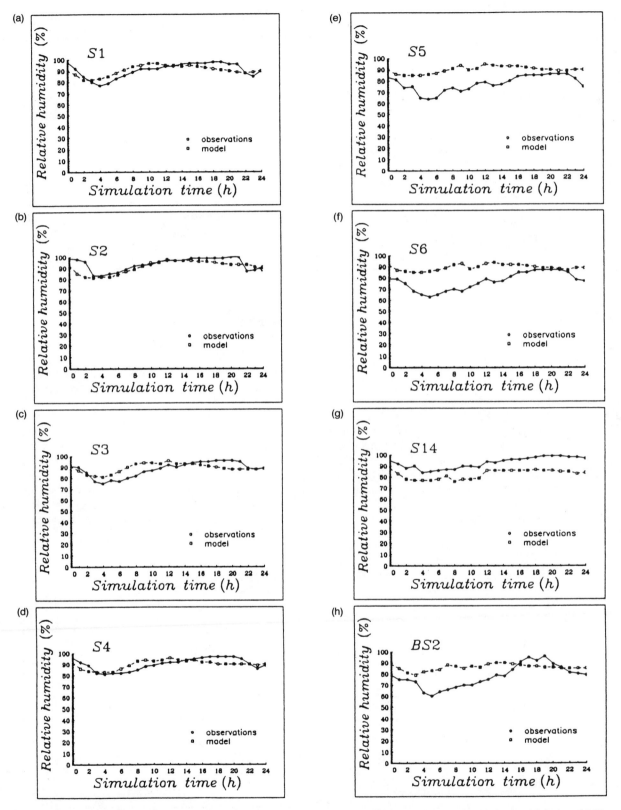

FIG. 10.19. (a)–(h) Model verification for diurnal variation of relative humidities at stations S1, S2, S3, S4, S5, S6, S14, and BS2. Model (dashed) and observation (full) curves are drawn.

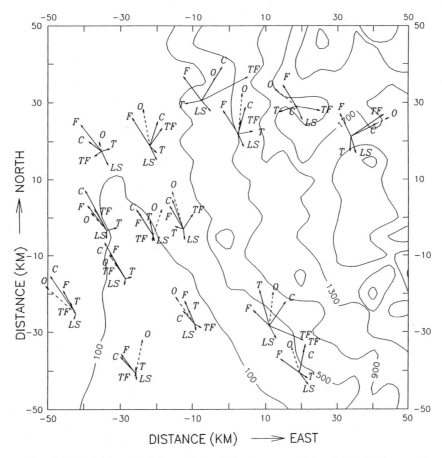

FIG. 10.20. Model domain 100 km × 100 km in the Sacramento Valley, California. Topographic contour interval is 400 m. Observed (**O**; dashed) wind vectors along with the control run result (**C**) for 1500 PST 27 June 1985 are plotted. Factor-separated wind vectors are topography (**T**), fluxes (**F**), topography-flux synergism (**TF**), and the residual large-scale (**LS**). The four contributions sum up to the control wind vector.

Figure 10.20 presents the factor-separated vectors in 15 locations where the observed wind was available. The observed wind was added for comparison with the control (**C**) result and is denoted by the **O** vector. One interesting result is that the synergistic **TF** contribution is dominant over the mountainous region, whereas at the Sacramento Valley the **F** contribution is dominant. This is not surprising since the **TF** represents the anabatic wind that is due to combined effect of thermal fluxes and sloping terrain. The **T** effect, however, represents only the mechanical effect and is relatively small at this time of the day. This method allows a better physical insight into the mechanisms responsible for the surface winds.

10.6. Conclusions

Our study shows that the high-resolution meso-β-scale simulations over Sacramento Valley, California, are highly sensitive to initial and boundary conditions. During summer, it was found that land-use variation can affect the Bowen ratio considerably, even to its doubling. It was also shown that initial surface moisture field from radiosondes is certainly insufficient and possibly inaccurate. During winter, the updating of the lateral BC was found crucial, but insufficient, for predicting the weak frontal passage at the early morning. Large spatial variability in moisture verification was found, and in both seasons subscale topographical features cause, as expected, significant deviations in the model verification.

Acknowledgments. We wish to thank the U.S. Army for the opportunity to use the unique WIND dataset for exploring model sensitivities and, in particular, R. Cionco, as well as the European Mesomet Panel. Special thanks are also due to Prof. Pearce for pursuing this subject.

APPENDIX

The PSU–NCAR Mesoscale Model

This meteorological model was developed by Anthes, Warner, and their colleagues at the National Center for Atmospheric Research (NCAR) and The Pennsylvania State University (PSU). The version of the PSU–NCAR model that was used in the studies is described in Anthes et al. (1987) and Hsie (1987). The development below follows that description.

a. Equations

The momentum equations are given by

$$\frac{\partial p^* u}{\partial t} = -m^2\left(\frac{\partial p^* u u m^{-1}}{\partial x} + \frac{\partial p^* v u m^{-1}}{\partial y}\right)$$

$$-\frac{\partial p^* u \dot{\sigma}}{\partial \sigma} - mp^*\left[\frac{RT_v}{(p^* + p_t \sigma^{-1})}\frac{\partial p^*}{\partial x} + \frac{\partial \phi}{\partial x}\right]$$

$$+ fp^* v + F_H u + F_V u$$

$$\frac{\partial p^* v}{\partial t} = -m^2\left(\frac{\partial p^* u v m^{-1}}{\partial x} + \frac{\partial p^* v v m^{-1}}{\partial y}\right)$$

$$-\frac{\partial p^* v \dot{\sigma}}{\partial \sigma} - mp^*\left[\frac{RT_v}{(p^* + p_t \sigma^{-1})}\frac{\partial p^*}{\partial y} + \frac{\partial \phi}{\partial y}\right]$$

$$- fp^* u + F_H v + F_V v,$$

where

$$p^* = p_s - p_t.$$

The continuity equation is

$$\frac{\partial p^*}{\partial t} = -m^2\left(\frac{\partial p^* u m^{-1}}{\partial x} + \frac{\partial p^* v m^{-1}}{\partial y}\right) - \frac{\partial p^* \dot{\sigma}}{\partial \sigma}.$$

The vertical integral of the continuity equation is used to compute the time variation of p^*:

$$\frac{\partial p^*}{\partial t} = -m^2 \int_0^1 \left(\frac{\partial p^* u m^{-1}}{\partial x} + \frac{\partial p^* v m^{-1}}{\partial y}\right) d\sigma'.$$

Then the vertical velocity in sigma coordinates is computed at each level in the model:

$$\dot{\sigma} = -\frac{1}{p^*}$$

$$\times \int_0^\sigma \left[\frac{\partial p^*}{\partial t} + m^2\left(\frac{\partial p^* u m^{-1}}{\partial x} + \frac{\partial p^* v m^{-1}}{\partial y}\right)\right] d\sigma'.$$

The thermodynamic equations are represented as follows:

$$\frac{\partial p^* T}{\partial t} = -m^2\left(\frac{\partial p^* u T m^{-1}}{\partial x} + \frac{\partial p^* v T m^{-1}}{\partial y}\right) - \frac{\partial p^* T \dot{\sigma}}{\partial \sigma}$$

$$- \frac{RT_v \omega}{c_{pm}(p^* + p_t \sigma^{-1})} + \frac{p^* Q}{c_{pm}} + F_{HT} + F_{VT},$$

where $\omega = p^* \dot{\sigma} + \sigma \dfrac{\partial p^*}{\partial t}$.

Also note that

$$\frac{dp^*}{dt} = \frac{\partial p^*}{\partial t} + m\left(u\frac{\partial p^*}{\partial x} + v\frac{\partial p^*}{\partial y}\right)$$

and $c_{pm} = c_p(1 + 0.8q_v)$.

The hydrostatic equation and virtual temperature are used to compute the geopotential heights:

$$\frac{\partial \phi}{\partial \ln(\phi + p_t/p^*)} = RT_v\left(1 + \frac{q_c + q_r}{1 + q_v}\right)^{-1},$$

where $T_v = T(1 + 0.608q_v)$.

The physics closures on the subgrid are first order, and detailed boundary parameterization is according to the modeling of Blackadar (Zhang and Anthes 1982). A grid-scaled diffusivity is also included for stability.

b. Definitions of symbols

The symbols used in the quoted equations are generally standard. Only those with meanings specific to the particular model concerned are defined here:

$$m\text{—map factor }\left[= \frac{\sin\psi_1}{\sin\varphi}\left(\frac{\tan\psi/2}{\tan\psi_1/2}\right)^{0.716}\right],$$

where $\psi_1 = 30°$, $\psi = 90° - \varphi$, and φ is latitude. Suffix s denotes surface and t denotes top of model atmosphere.

c. Solution technique

The finite-difference methods use an Arakawa B staggered grid and approximately conserve mass, mo-

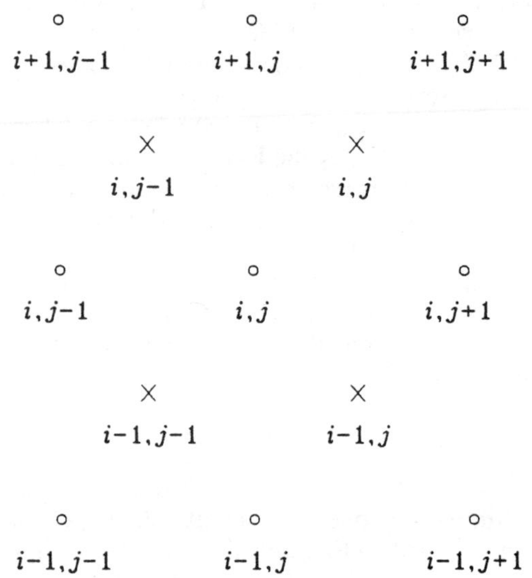

FIG. 10A.1. Depiction of dot points and offset cross points.

mentum, and total energy. The $p*u$ and $p*v$ variables are calculated at so-called dot points and all other variables are calculated at offset cross points, using $y(i)$, $x(j)$ indexing as shown in Fig. 10A.1.

A typical finite-difference equation is represented by the u component of the equations of motion:

$$\frac{\partial p*u}{\partial t} \approx -m^2\left[\left(\bar{u}^x\,\overline{\frac{\overline{p*u}^y}{m}}^x\right)_x + \left(\bar{u}^x\,\overline{\frac{\overline{p*v}^x}{m}}^y\right)_y\right]$$

$$-\frac{\delta\bar{\sigma}^{xy}\overline{\overline{p*u}}^\sigma}{\delta\sigma} - m\overline{p*}^{xy}\overline{\phi_x}^y$$

$$-\frac{mR\overline{T_v}^{xy}}{(1 + p_t/\overline{p*}^{xy}\sigma)}\,\overline{p*_x}^y + fp*v + F_H u + F_V u.$$

Subscript coordinates represent two-point derivatives, overbars represent two-point averages per coordinate, and over brackets represent four-point operations:

$$\alpha_x = \frac{1}{\Delta s}[\alpha(i, j + \tfrac{1}{2}) - \alpha(i, j - \tfrac{1}{2})]$$

$$\bar{\alpha}^x = \frac{1}{2}[\alpha(i, j + \tfrac{1}{2}) + \alpha(i, j - \tfrac{1}{2})]$$

$$\overline{\overline{\alpha^x}} = \frac{1}{4}[\alpha(i, j + 1) + 2\alpha(i, j) + \alpha(i, j - 1)].$$

The accuracy of the model was found to be sensitive to the form of the hydrostatic equation used. The best results were found by integrating the hydrostatic equation above to find ϕ at velocity levels.

Chapter 11

Simulations of Project WIND Cases with RAMS

ROBERT L. WALKO AND ROGER A. PIELKE

Department of Atmospheric Science, Colorado State University, Fort Collins, Colorado

11.1. Introduction

The Regional Atmospheric Modeling System (RAMS) is a general purpose numerical code for simulating a wide variety of atmospheric flows. Versatility is one of the major goals in its design; the code is structured such that by setting a number of parameters to appropriate values from a large menu of options the model can be applied to nearly any flow situation. RAMS has been used to model flows at scales ranging from hemispheric circulations to submicroscale eddies induced by flow over buildings (e.g., Tremback et al. 1986; Cotton et al. 1988; Tripoli and Cotton 1989a,b; Schmidt and Cotton 1990; Cram et al. 1992a,b; Lee et al. 1993; Pielke et al. 1992). It contains a full complement of subgrid, microphysical, radiative, convective, and surface-layer parameterizations, along with a prognostic model for soil and vegetation temperature and moisture. Other options include two-way interactive grid nesting, terrain-following coordinates, several lateral and upper boundary conditions, and stretched vertical coordinates.

RAMS is used primarily as a research, rather than operational, model. The distinction is important because it implies, as stated above, that the model configuration is not fixed or hard wired but flexible. Operational models are optimized and tuned for their specific application, whereas RAMS requires experimentation to achieve optimal performance for any new situation. RAMS has a comprehensive manual to guide the user through the process of configuring the model, but it is impossible for it to be any more than a guide; setting parameters is in part an art form continually learned by even the most experienced users. If 10 experienced users of RAMS are given the task of carrying out a simulation for a specified location and time, they will invariably do so with 10 different combinations of 1) horizontal and vertical extent of overall domain; 2) number of vertical levels; 3) number and sizes of nested grids; 4) vertical grid stretching; 5) smoothness of topography; 6) choice of subgrid, cumulus, microphysical, soil, and vegetation parameterizations; 7) initial soil temperature and moisture; and 8) details of atmospheric initialization, among many others, and they

will obtain 10 quantitatively different results. This is in no way meant as any sort of apology; indeed, we are pleased with the model's performance for this and other applications. Rather, these comments are intended to emphasize that the results presented are, to a large extent, arbitrary in their details, being among many possible outcomes produced by RAMS. Such flexibility in setting up the model also provides guidance as to the limits of the uncertainty of the model results.

11.2. RAMS application

For the present application, we configured RAMS in the following way. Three nested grids were used, the smallest of which (grid 3) was a mesh of 40×40 points and 5-km spacing in the horizontal, which coincides with the 200 km $\times$ 200 km Project WIND domain. The telescoping middle and outer grids (grids 2 and 1), which had 20- and 80-km horizontal grid spacings, respectively, provided a large domain surrounding grid 3. Thus, grid 3 had high-quality two-way communication at its lateral boundaries, allowing it to exchange information with the surrounding atmospheric fields. The open lateral boundaries of grid 1, where less-reliable boundary conditions must be applied, are well over 1000 km removed from the Project WIND domain, and thus have little effect on it during the period of the simulation. Figure 11.1a shows the domains of the three grids, and Fig. 11.1b shows the surface topography on grid 3. Thirty-three vertical levels were used on each grid, having a vertical spacing of 80 m near the ground, stretching exponentially to 1 km at 8-km altitude, and remaining a constant 1 km above. This places the model top near 18 km. The model simulation is begun at 1200 UTC (0400 PST) for both Project WIND phases, which is a time when rawinsonde data are available. Model fields are initialized from a combination of rawinsonde and surface observations, and the National Meteorological Center (NMC) 2.5° global gridded analysis. These data are objectively analyzed on isentropic surfaces from which they are interpolated to the RAMS grid. Similar analyses are prepared for the subsequent three observational

(a)

(b)

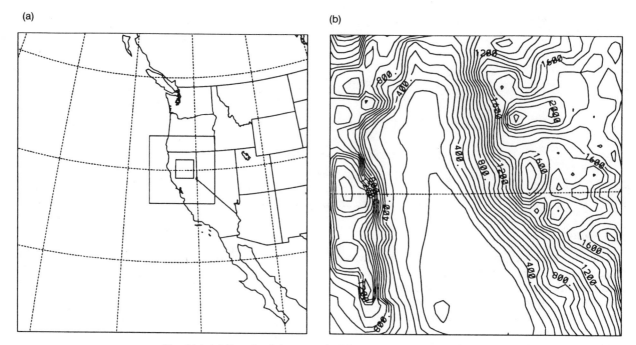

FIG. 11.1. (a) Domain of three nested grids. (b) Topography for grid 3.

periods (spaced at 12-h intervals) and are used to nudge the lateral boundary region of grid 1 during the simulation. No nudging is done in the interior of grid 1 nor in any part of grids 2 or 3. For the phase I simulation, atmospheric moisture is carried in vapor form only, whereas for phase II, full microphysics was employed.

RAMS is equipped with a prognostic soil model that predicts temperature and moisture within multiple layers of the soil. Usually about 10 levels are specified spanning the first half to 1 m of depth. Each modeled soil layer is a reservoir for storing heat and water, and these quantities are transferred between the layers through diffusion. Exchanges of sensible and latent heat

(a)

(b)

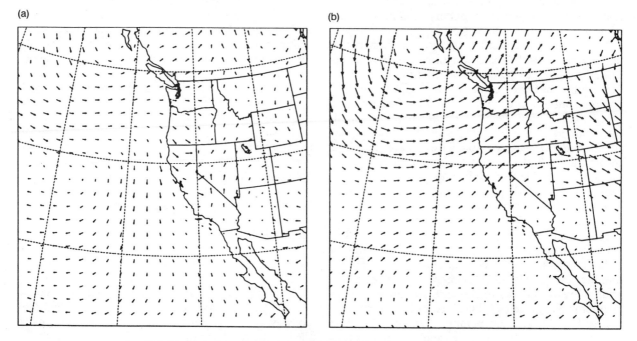

FIG. 11.2. Modeled grid 1 wind fields at 1500 PST for phase I: (a) surface and (b) 5400 m.

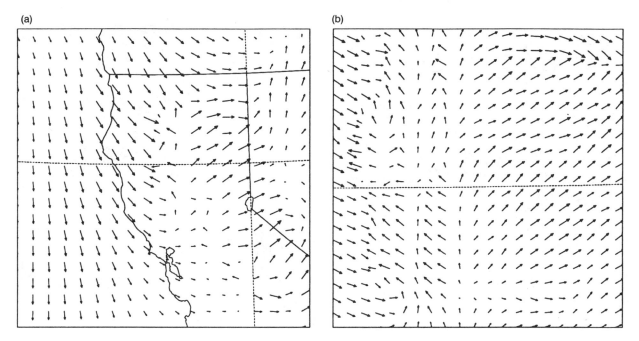

FIG. 11.3. Modeled surface wind fields at 1500 PST for phase I: (a) grid 2 and (b) grid 3.

(moisture) between the top soil layer and the atmosphere are modeled using boundary layer similarity theory. The soil surface energy budget includes shortwave and longwave radiative transfer.

Numerous experiments with RAMS have demonstrated that initial values of moisture in the soil are highly important in influencing the surface energy budget (e.g., Ookouchi et al. 1984). The primary reason is that soil moisture strongly affects the Bowen ratio, thus regulating the sensible heat flux over a possible wide range of values. While the initial soil temperature is also of some importance, the amount of error in thermal energy storage in the soil due to a poor estimate of temperature has far less effect on the overlying atmosphere than an error in latent heat energy storage from a poor moisture estimate, given reasonably intelligent estimates on each. Unfortunately, soil moisture content is usually a poorly known quantity, not routinely measured, and atmospheric models must rely on crude initial estimates of soil moisture. The user of RAMS is free (and encouraged) to experiment with soil moisture for any given application, using guidance where possible from any available information such as the *Weekly Crop Bulletin.* There is, however, a default initialization procedure that equates a soil relative humidity to the initial atmospheric surface relative humidity based on the assumption that moist soil may be responsible for moist surface air.

a. Model results—Phase I of WIND

This default was initially used for the Project WIND simulations with unfavorable results. Some regions of

the domain began with very high relative humidity values. This is partly due to the initialization time (1200 UTC or 0400 PST), which is when relative humidity values are normally highest, but also probably due to erroneously low temperatures in the initial data. The high soil moisture content resulting from the default initialization procedure caused large amounts of latent heat to be released at the expense of sensible heat, and the model surface temperatures remained low all day.

This discrepancy was noted upon comparing RAMS results to observations from Project WIND, and the simulation was subsequently rerun with a lower soil moisture. Justification for the lower moisture was obtained from the *Weekly Crop Bulletin,* which indicated the entire region to be abnormally dry for the period.

Results from the drier soil run provided the figures presented in this paper. Figure 11.2 shows the wind vectors near the surface and at 5400 m MSL at 1500 PST, which is 11 h after model initialization or 6 h after the beginning of the Project WIND comparison period. A weak cyclone in the northern region of the domain is the most prominent feature at the higher level. The Project WIND domain lies under the southeastern side of the cyclone, giving the region southwesterly winds around 13 m s^{-1} at that level. Near the surface, winds are weaker and highly variable in direction, with a maximum of 11 m s^{-1} over mountainous terrain in California. Figure 11.3 shows the near-surface winds at the same time for grids 2 and 3. Winds over the ocean on grid 2 are northerly or northwesterly and have an onshore component until the crest of the Coast Range is reached. Just east of the crest, winds have an easterly component, while still farther east, winds are

(a)

(b)

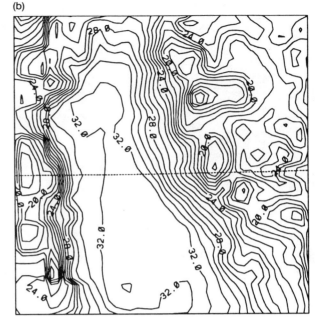

(c)

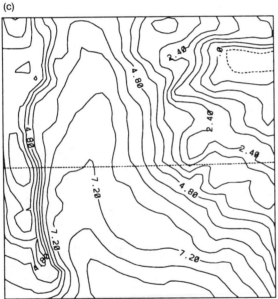

FIG. 11.4. Modeled surface conditions on grid 3 at 1500 PST for phase I: (a) streamlines, (b) temperature, and (c) dewpoint.

again westerly as they ascend the Sierra Nevada. This pattern is seen in greater detail in grid 3. A principal cause of these wind patterns is solenoidal forcing induced by solar heating of the sloping terrain, which is near a maximum at this hour of the afternoon. Winds on grid 3 also contain a southerly component in most areas.

Figure 11.4a presents a streamline plot of the modeled low-level grid 3 winds, which further emphasizes the regions of convergence and divergence. Figures 11.4b and 11.4c show the horizontal distributions of modeled temperature and dewpoint on grid 3. Vertical cross sections of temperature, potential temperature,

dewpoint, and vertical velocity through the center latitude of grid 3 are presented in Figs. 11.5a–d. Temperatures at the valley floor reach 33°C, with dewpoint temperatures reaching around 8°C. Figure 11.5b reveals that the model simulated a fairly deep boundary layer, whose depth exceeds 2 km over the valley. The vertical velocity plot shows a tendency for thermal convection within the boundary layer, which has a strong effect on instantaneous surface winds, although a 5-km model grid spacing is far too coarse to allow convection to freely develop.

Figure 11.6 shows surface wind vectors, temperature, and dewpoint for the phase I simulation at 0300 PST,

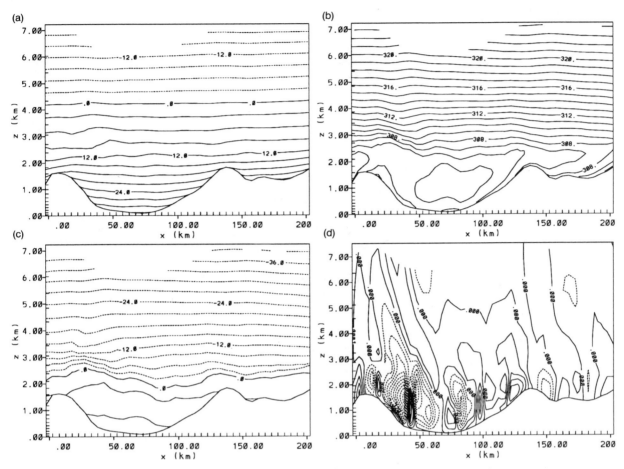

FIG. 11.5. Vertical cross sections of modeled conditions on grid 3 at 1500 PST for phase I: (a) temperature,
(b) potential temperature, (c) dewpoint, and (d) vertical velocity.

or 12 h after the previous figures. Figure 11.7 shows vertical cross sections of temperature, potential temperature, and dewpoint for the same time. Low-level nocturnal cooling has led to downslope flows into the valley, with drainage flow southward along the valley floor. Surface winds over the higher terrain in the east are mostly southerly. Dewpoint temperatures in the valley have decreased slightly from the previous 12 h.

We surmise that initial soil moisture in RAMS was still too high for the phase I simulation. Modeled dewpoint values are generally several degrees too high in this simulation, particularly in the early hours. Figure 11.8 shows time series of dewpoints from four different surface stations and the corresponding RAMS values interpolated to those locations. The four stations were selected from a total of 14 that reported moisture to show the variety of comparison with RAMS. (Note that in Fig. 11.8 and all other time-dependent plots, the time variable denotes hours from the beginning of the Project WIND comparison time. Thus, for phase I the time of 0 h occurs at 0900 PST.) At station S8 (Fig. 11.8a), the observed dewpoint was consistently several degrees below RAMS throughout the period.

Similar comparisons occurred at stations S2, S7, and S9 (not shown). Station S3 (Fig. 11.8b) began with RAMS several degrees moister and ended with RAMS somewhat drier. This trend also occurred at stations S1, S4, S5, and BS2. Stations S13 (Fig. 11.8c), S12, and S14 are similar to Fig. 11.8b, except that observed dewpoints were lower for the last 10 h.

Station S10 (Fig. 11.8d) began unusually moist compared to other stations, exceeding even RAMS's moisture, subsequently drying and remoistening. RAMS dewpoints showed very little variability from one station to another, while observations were highly variable between stations. Two likely explanations for this behavior are 1) RAMS values are essentially averaged over grid cells 5 km × 5 km × 80 m, and thus are incapable of representing variations on smaller scales that are detected by instruments, and 2) the actual ground surface is characterized by small-scale variations in vegetation type, surface roughness, soil type, soil moisture, and other properties (e.g., Avissar and Pielke 1989; André et al. 1989) that influence local meteorology, but this finescale representation is not possible on a 5-km model grid.

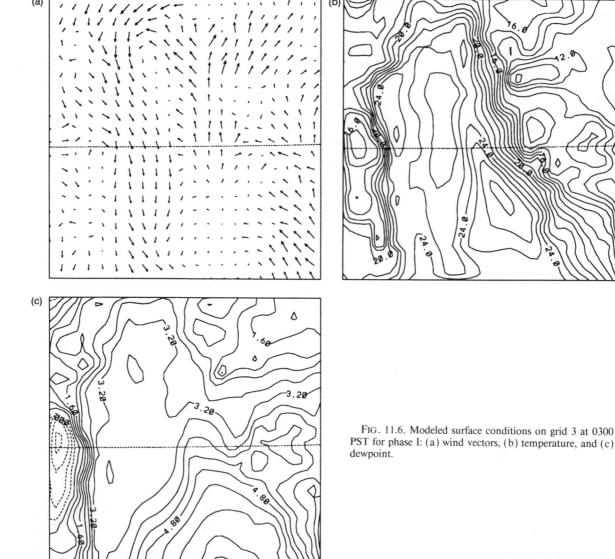

FIG. 11.6. Modeled surface conditions on grid 3 at 0300 PST for phase I: (a) wind vectors, (b) temperature, and (c) dewpoint.

Temperature comparisons between RAMS and Project WIND data fall into three broad types, shown in Fig. 11.9. Station S2 (Fig. 11.9a) and similarly stations S5, S7, S9, and S11 begin several degrees colder than observed temperatures but warm up to nearly the same peak temperature by afternoon. Subsequently, the temperatures match rather well. Station S12 (Fig. 11.9b) and similarly most other stations fall 2° or 3° short of reaching the observed peak temperature and subsequently do not become as cold the following night. The two cases that differ from all others are stations C2 (Fig. 11.9c) and C3, for which RAMS temperature varies over only a range of 6°C, whereas the observed temperature varies by more than three times that

amount. These are higher elevation stations, and apparently influenced by upper-level winds in RAMS, which were somewhat stronger than observed (Fig. 11.10a). Figure 11.9d shows observed and modeled solar radiation incident at the ground at station S3. The comparison is excellent, as it was at all other stations that reported radiation. Occasionally, however, brief periods of cloudiness reduced the incoming solar at some stations, whereas the RAMS simulations for phase I neglected radiative effects of clouds.

Figure 11.10 shows wind speed comparisons at four stations, and Fig. 11.11 shows the corresponding wind direction comparisons. Figure 11.10a, as already noted, shows RAMS to have a stronger surface wind at station

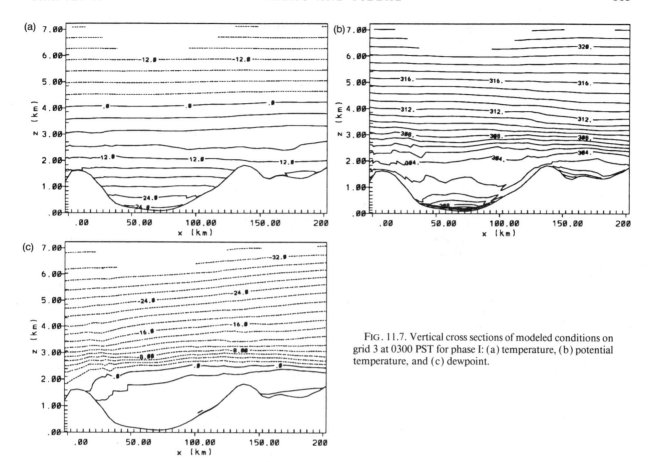

FIG. 11.7. Vertical cross sections of modeled conditions on grid 3 at 0300 PST for phase I: (a) temperature, (b) potential temperature, and (c) dewpoint.

C2 than that observed over most of the period. All other stations had reasonably comparable mean values, however. At many stations, including S8 (Fig. 11.10b), RAMS winds were somewhat weaker than observations in the early part of the simulation when observed winds peak, and stronger than observed later. In other cases, such as station S5 (Fig. 11.10c), observed winds lack a peak in the afternoon and do not vary appreciably over the period. Figure 11.10d shows a fairly unusual case (station S11) where observed winds remained strong through the early morning hours and exceeded RAMS speeds. In spite of differences in detail, RAMS speeds were generally reasonably close in magnitude to the observations.

Wind directions generally compared better than speeds. Figure 11.11c (station S5) is one of several examples where directions match quite well. Figure 11.11b (station S8) appears to match poorly during a several-hour period, but the observed and modeled directions are actually reasonably close, being on opposite sides of the 0° azimuth. Figure 11.11d (station S11) does more poorly during roughly the same period but matches very well at other times. Figure 11.11a (station C2) is one of the poorest direction comparisons, although the observed winds were generally quite weak and thus did not have a robust direction.

Comparisons of upper-air profiles between observations and RAMS are made for stations WSC and WSF. (Stations WSO and C32 had the most missing data, and many data from station WSH appear questionable.) Times are selected as late afternoon (1700 PST) near the warmest time of day, and 12 h later at 0500 PST. Figure 11.12 shows temperature profiles at both times for both stations. Reasonably good agreement is found particularly in lapse rate, with RAMS within 3° or 4°C of observed temperatures at all levels. Differences as large as 6°C were found at low levels at certain other times and stations.

Dewpoint profiles corresponding to temperatures of Fig. 11.12 are shown in Fig. 11.13. RAMS is, on the average, moister than the observed atmosphere, although agreement is good at station WSC at 0500 PST (Fig. 11.13b). As with surface conditions, RAMS exhibits less variability from station to station than observations in both temperature and moisture.

Wind speeds for both stations and both times are shown in Fig. 11.14. Station WSC at 1700 PST exhibits widely varying wind speeds as a function of height

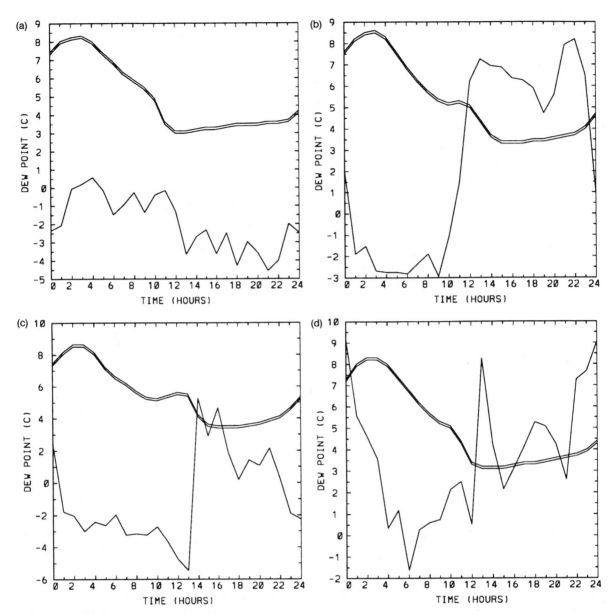

FIG. 11.8. Comparison of modeled (double line in this and subsequent plots) and observed (single line in this and subsequent plots) surface dewpoints for phase I as a function of time: (a) station S8, (b) station S3, (c) station S13, and (d) station S10. In this and subsequent phase I time series plots, the time of 0 h corresponds to 0900 PST.

(which may be questionable), while RAMS produces a much smoother profile. A general trend of increasing observed speed with height is discernible, which agrees well with RAMS. Station WSF (Fig. 11.14c) at this time matches well at all levels between RAMS and observations. At 0500, agreement is reasonably good for both stations, although RAMS again has smoother vertical profile than observations.

Corresponding wind directions are presented in Fig. 11.15. Agreement is quite good over nearly all levels

for station WSC at 1700 PST (Fig. 11.15a). The largest discrepancies occur at low levels where winds are relatively weak and wind direction is thus likely to be highly variable. Station WSF at 1700 PST (Fig. 11.15c) exhibits rather poor agreement in wind direction below 1500 m (although speed is weak there) but shows excellent agreement above 1500 m. At 0500 PST, there is very good agreement except where winds are weak. (The recorded observed wind direction is set to zero when winds are very weak.)

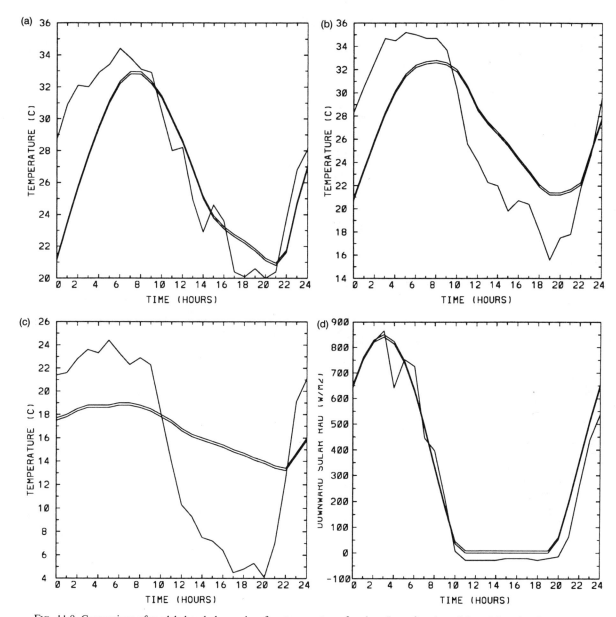

FIG. 11.9. Comparison of modeled and observed surface temperatures for phase I as a function of time: (a) station S2, (b) station S12, (c) station C2. (d) Comparison of modeled and observed downward solar radiation flux at station S3.

b. Model results—Phase II of WIND

Phase II is characterized by stronger synoptic forcing, thus relying more on influences entering the domain from outside than from heating of local terrain features. In this case, RAMS's ability to nest the Project WIND domain inside larger grids was of particular benefit. The coarse grid was large enough to contain in its initial fields all large systems close enough to influence the Project WIND domain within the period of simulation. Soil was again initialized fairly dry (as indicated by the

Weekly Crop Bulletin), although this is of less consequence than in the summer case where surface heating is of paramount importance.

Figure 11.16 shows the modeled low-level and 5400-m winds at 1000 PST, the beginning time of the Project WIND comparison. A strong westerly jet of 50 m s^{-1} at the higher level lies off the coast, with diffluent winds over California. The Project WIND domain has westerly winds of 24 m s^{-1} at this level. Surface winds contain a similar westerly jet (reaching 20 m s^{-1}) over the ocean, but within the Project WIND domain, sur-

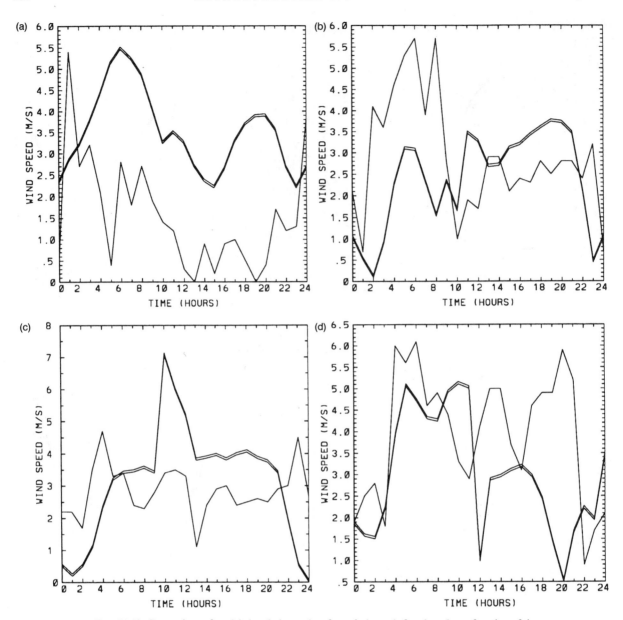

FIG. 11.10. Comparison of modeled and observed surface wind speeds for phase I as a function of time:
(a) station C2, (b) station S8, (c) station S5, and (d) station S11.

face winds are relatively weak at this time. Figure 11.17a shows the grid 3 surface winds, which reach 12 m s^{-1} over higher terrain but are generally less than half that value in the valley. Winds are mostly south to southeasterly flowing up the valley floor, but tend to be more southwesterly over higher terrain. Figures 11.17b and 11.17c show surface distributions of temperature and dewpoint for grid 3, and Figs. 11.18a–e show vertical profiles of wind vectors, temperature, potential temperature, dewpoint, and vertical velocity. Fairly strong winds are simulated over the Coast Range

and above, leading to the characteristic mountain wave pattern evident in Fig. 11.18e.

Approximately 18 h later, a wind surge was observed at nearly all stations in the Project WIND domain. Figures 11.19a–c show surface winds, temperature, and dewpoint for this time (0400 PST), and Figs. 11.20a–c show vertical cross sections of wind, temperature, and dewpoint. Temperatures over the entire domain have warmed from 18 h earlier, with the average temperature increasing by about 3°C. Dewpoints in the valley have fallen only slightly, but over the higher terrain the air

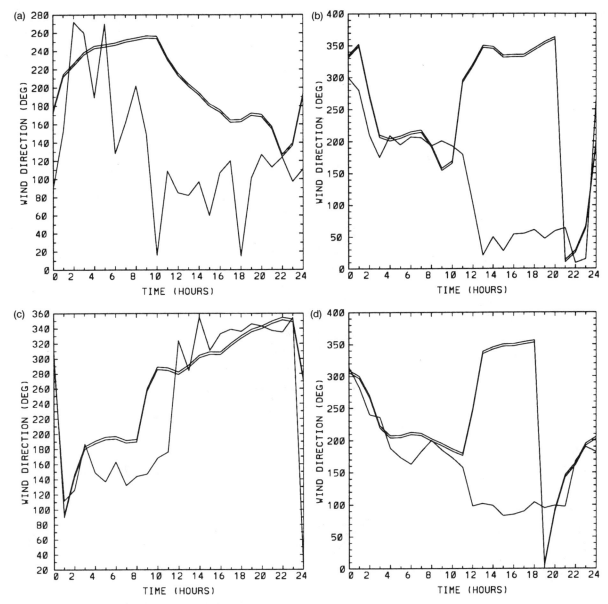

FIG. 11.11. Comparison of modeled and observed surface wind directions for phase I as a function of time: (a) station C2, (b) station S8, (c) station S5, and (d) station S11.

has dried considerably. Wind vectors appear quite similar to 18 h earlier, but magnitudes are approximately double what they were. Thus, the model indicates that the wind surge is a fairly widespread phenomenon. Low-level wind vectors for grid 2 are plotted in Fig. 11.21, showing that strong winds occur well beyond the confines of grid 3. This is a further indication that successful simulation of the wind surge in grid 3 required communication from outside the grid 3 region, given that the simulation was begun at least 18 h beforehand.

Comparisons between observed and modeled surface temperatures at several station locations are shown in Fig. 11.22. (Note that in Fig. 11.22 and all other time-dependent phase II plots, the time of 0 h occurs at 1000 PST.) At most stations (e.g., station S3 in Fig. 11.22a), surface temperatures in RAMS warmed slowly during the first few hours, remained fairly constant during the nighttime hours, and continued a warming trend following sunrise the next morning. Also at most stations (e.g., station S3), observed temperatures warmed more rapidly the first morning, cooled during the night to

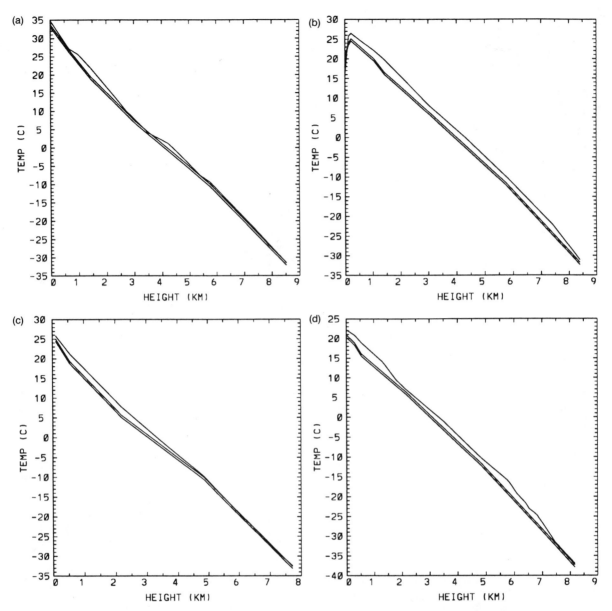

FIG. 11.12. Comparison of modeled and observed temperature profiles for phase I: (a) station WSC at 1700 PST,
(b) station WSC at 0500 PST, (c) station WSF at 1700 PST, and (d) station WSF at 0500 PST.

near or slightly below RAMS temperatures, and rose the following morning. Figure 11.22b illustrates temperatures at station C3, which is at a higher elevation and has colder temperatures. Here, RAMS warms by 4°C during the night due to stronger than observed surface winds (Fig. 11.24c), while surface temperatures warm only slightly. This is the most extreme case of RAMS exceeding observed surface temperatures. The opposite extreme occurs at station S10 (Fig. 11.22c), where the observed temperature is higher than any

other station and exceeds RAMS temperatures by up to 6°C in the afternoon.

The relatively cool temperatures in RAMS early in the simulation are primarily caused by the model's simulation of cloud cover at several locations. This caused strong attenuation of downward solar radiation, as typified at station S3 (Fig. 11.22d). Of the few stations that reported incoming solar radiation, station S14 matches well, due to the lack of cloud cover in RAMS at that location. In spite of attenuated radiation,

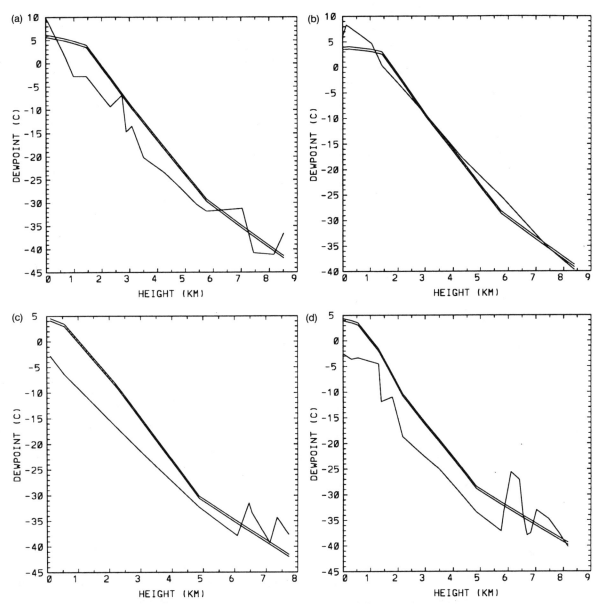

FIG. 11.13. Comparison of modeled and observed dewpoint profiles for phase I: (a) station WSC at 1700 PST, (b) station WSC at 0500 PST, (c) station WSF at 1700 PST, and (d) station WSF at 0500 PST.

RAMS temperatures were generally closer to observed values in phase II than in phase I. This is probably due to the strong influence of synoptic forcing relative to the surface forcing, coupled with the uncertainty in soil moisture.

Dewpoint values were also better matched between RAMS and observations in phase II. Differences never exceeded 5°C at any station at any time, and were in most cases within 2°C. At all surface stations, modeled dewpoints trended upward by 1° or 2°C during the

simulation period, while observed values exhibited a range of behavior. Figure 11.23a (station S3) typifies several stations at which observed dewpoints reached close to 12°C early in the period, and dropped to around 10°C later. Most stations not following this pattern behaved similarly to station S4 (Fig. 11.23b), remaining around 7°–8° through the period.

Comparisons between observed and modeled wind speeds are shown in Fig. 11.24. At all station locations, RAMS produced a peak wind speed at 18–19 h into

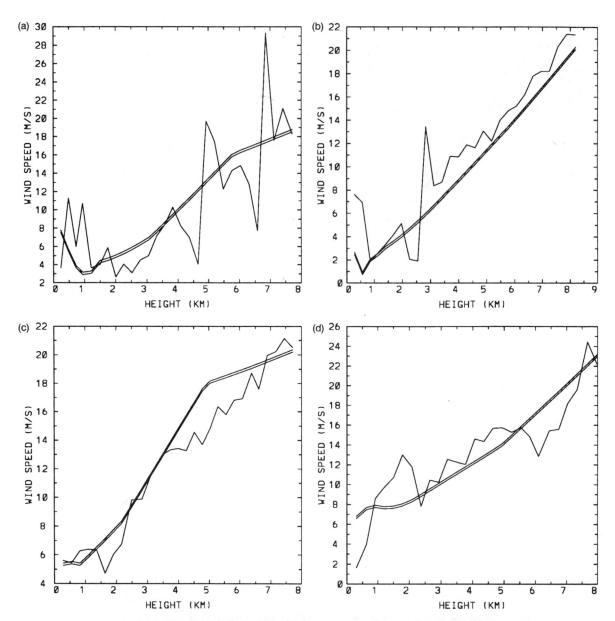

FIG. 11.14. Comparison of modeled and observed wind speed profiles for phase I: (a) station WSC at 1700 PST, (b) station WSC at 0500 PST, (c) station WSF at 1700 PST, and (d) station WSF at 0500 PST.

the simulation (0400–0500 PST). This matches quite well a peak in wind speed observed at nearly every station at nearly the same time. Comparisons in wind speed magnitudes showed a variety of behavior. Around a third of the stations recorded winds approximately twice the modeled value, as at station S3 (Fig. 11.24a). Another third of the stations were close to matching wind speeds with RAMS, as at station S14 (Fig. 11.24b). A few stations had RAMS wind speeds exceeding observed values—the most extreme example of which is the more elevated station C3 (Fig. 11.24c),

which recorded no wind surge, while RAMS still simulated a strong one.

For most stations and times, comparisons of surface wind direction were excellent. At station S8 (Fig. 11.25a), for example, modeled and observed winds were nearly always within 10° and had equal mean directions. As for other fields, RAMS has little variation from one station to another, but observed conditions vary more. A few stations had wind directions significantly different from others (and from RAMS), the most extreme case of which is probably station S4 (Fig.

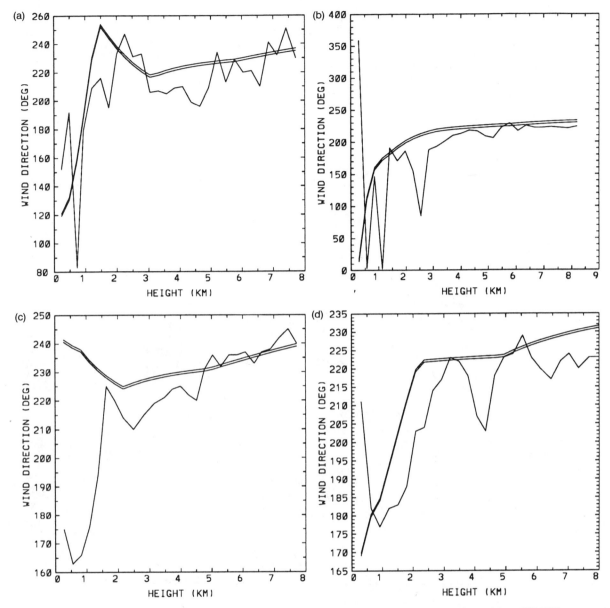

FIG. 11.15. Comparison of modeled and observed wind direction profiles for phase I: (a) station WSC at 1700 PST, (b) station WSC at 0500 PST, (c) station WSF at 1700 PST, and (d) station WSF at 0500 PST.

11.25b), where directions once reached a 50° difference.

Upper-air observational datasets are more complete for stations 2 and 4 than for the other stations, as was the case in phase I, so again those stations are chosen for comparison. Even at station WSC, data are missing from 0200 and 0400 PST, so we present comparisons for the time 0600 toward the end of the simulation period, plus 12 h earlier at 1800. Figure 11.26 shows profiles of observed and modeled wind speeds for both stations and both times. In all cases, agreement is good

in terms of mean (vertically averaged) wind speed and in the general trend that wind speeds increase with height. The degree of trend is sometimes a poor match, however, as for station WSF at 1800 PST (Fig. 11.26c), where observed winds decrease somewhat above 4 km, becoming considerably less than the modeled value by 8 km. An opposite, though less extreme case occurs at station WSC at 0600 PST (Fig. 11.26b), where modeled winds are stronger than observed in the lowest 3 km, but reasonably close to observations above. Station WSC at 1800 PST (Fig. 11.26a) and station WSF at

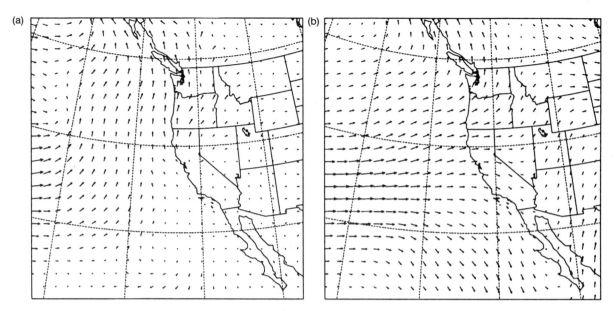

FIG. 11.16. Modeled grid 1 wind fields at 1000 PST for phase II: (a) surface and (b) 5400 m.

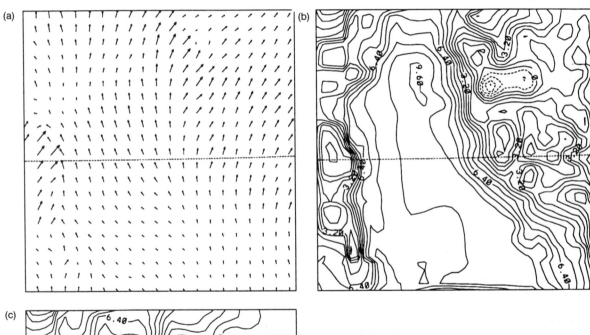

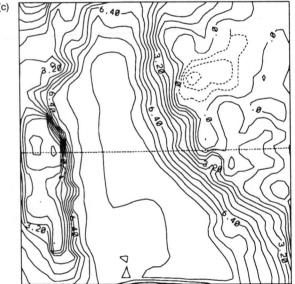

FIG. 11.17. Modeled surface conditions on grid 3 at 1000 PST for phase II: (a) wind vectors, (b) temperature, and (c) dewpoint.

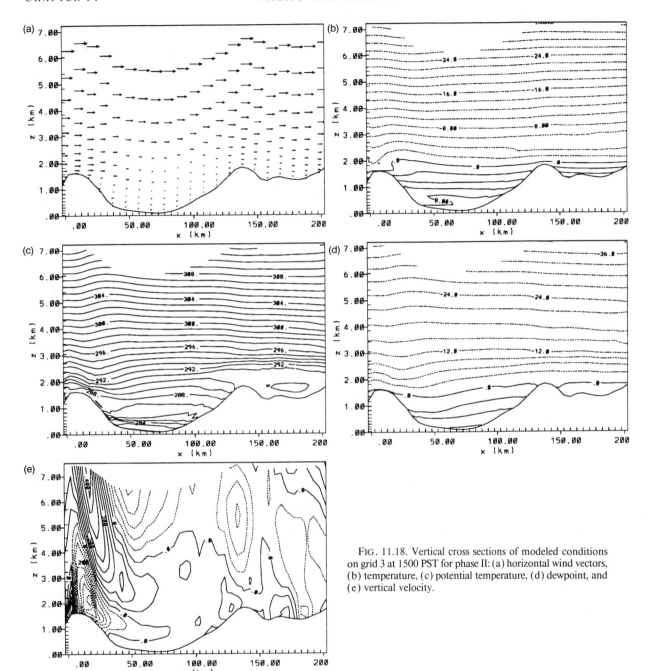

FIG. 11.18. Vertical cross sections of modeled conditions on grid 3 at 1500 PST for phase II: (a) horizontal wind vectors, (b) temperature, (c) potential temperature, (d) dewpoint, and (e) vertical velocity.

0600 PST (Fig. 11.26d) match reasonably closely between model and observations. In all cases, RAMS produces a much smoother vertical profile of wind speed than observed.

Upper-air wind directions are presented in Fig. 11.27 corresponding to the speeds in Fig. 11.26. Directions match very well for station WSC both at 1800 and 0600 PST (Figs. 11.27a and 11.27b). At station WSF, wind directions in RAMS are shifted systematically from observed directions by approximately 20° at most levels.

11.3. Conclusions

Two observational periods of the Project WIND field experiment were simulated using RAMS. The model performed a reasonably realistic simulation of both the summer and winter cases; however, the comparison of the modeling results with the observations illustrated two major limitations of the simulation of these meteorological events. For the summer case, the sensitivity of the results to soil moisture was convincingly demonstrated. Since information on this input information

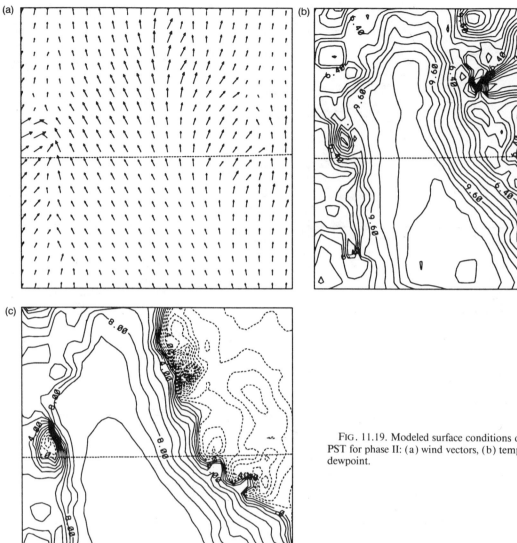

FIG. 11.19. Modeled surface conditions on grid 3 at 0400 PST for phase II: (a) wind vectors, (b) temperature, and (c) dewpoint.

is almost always lacking, this limitation provides bounds on our ability to obtain the most accurate quantitative results. The winter case documented the importance of large-scale synoptic information that is communicated from the coarser model grids to the fine grid that covers the Project WIND domain by means of interactive grid nesting. This communication was essential for the successful simulation of the observed wind surge.

Acknowledgments. Datasets used to initialize the RAMS simulations were obtained from the National Center for Atmospheric Research, which is supported by the National Science Foundation. Observational data from the Project WIND field experiment, which were used for comparison with RAMS results, consisted of all SAMS, Campbell, and GMD upper-air data. Project WIND was conducted by United States Army Laboratories and the United States Department of Agriculture Forest Service.

This work was supported through funding provided by the U.S. Army through a grant awarded to the Department of Meteorology at the University of Reading, United Kingdom and under National Science Foundation Grant ATM-8915265. The manuscript was ably completed by Ms. Dallas McDonald.

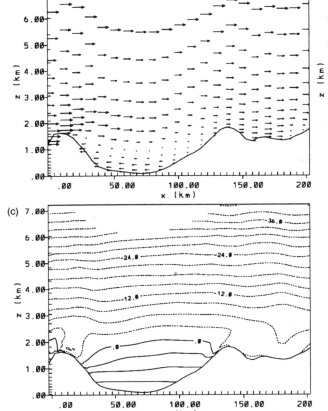

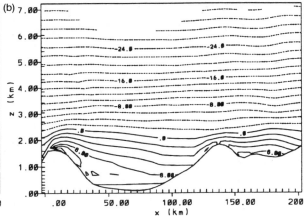

FIG. 11.20. Vertical cross sections of modeled conditions on grid 3 at 0400 PST for phase II: (a) horizontal wind vectors, (b) temperature, and (c) dewpoint.

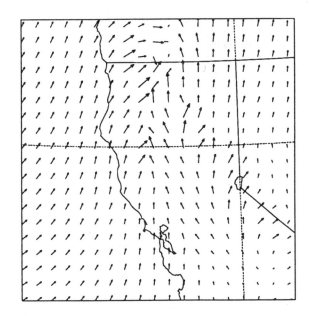

FIG. 11.21. Modeled surface wind vectors on grid 2 at 0400 PST for phase II.

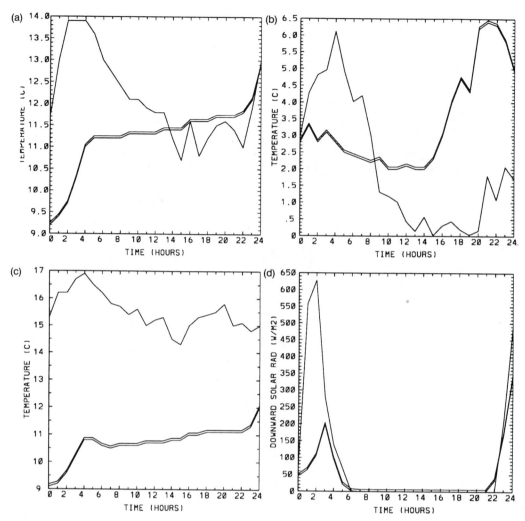

FIG. 11.22. Comparison of modeled and observed surface temperatures for phase II as a function of time: (a) station S3, (b) station C3, (c) station S10. (d) Comparison of modeled and observed downward solar radiation flux at station S3. In this and subsequent time series plots for phase II, a time of 0 h corresponds to 1000 PST.

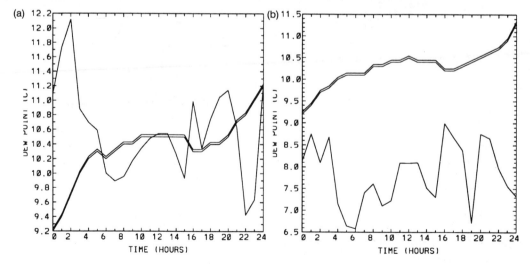

FIG. 11.23. Comparison of modeled and observed surface dewpoints for phase II as a function of time: (a) station S3 and (b) station S4.

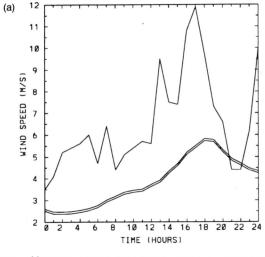

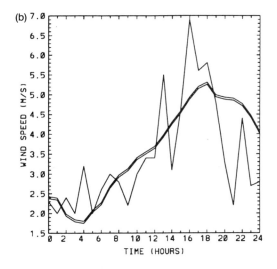

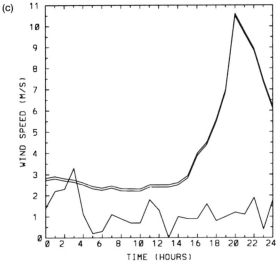

FIG. 11.24. Comparison of modeled and observed surface wind speeds for phase II as a function of time: (a) station S3, (b) station S14, and (c) station C3.

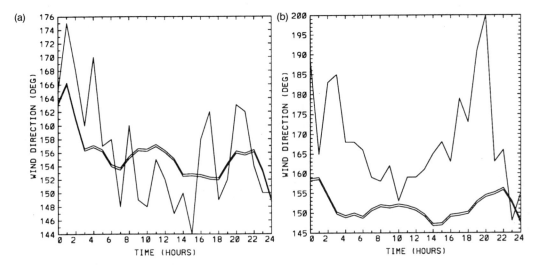

FIG. 11.25. Comparison of modeled and observed surface wind directions for phase II as a function of time: (a) station S8 and (b) station S4.

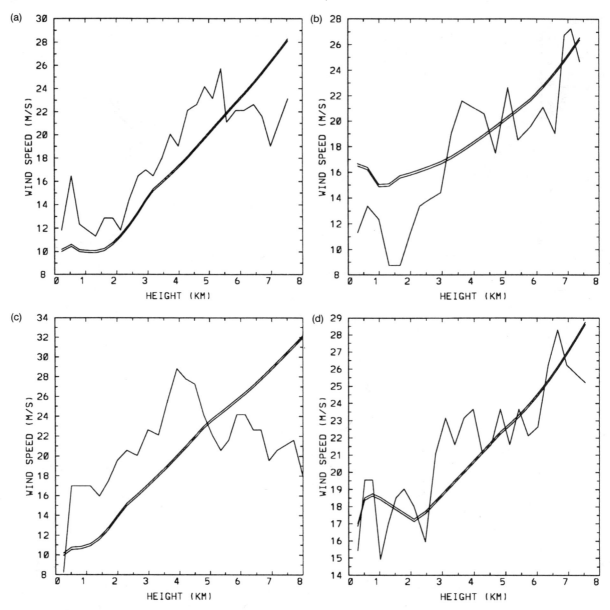

FIG. 11.26. Comparison of modeled and observed wind speed profiles for phase II: (a) station WSC at 1800 PST, (b) station WSC at 0600 PST, (c) station WSF at 1800 PST, and (d) station WSF at 0600 PST.

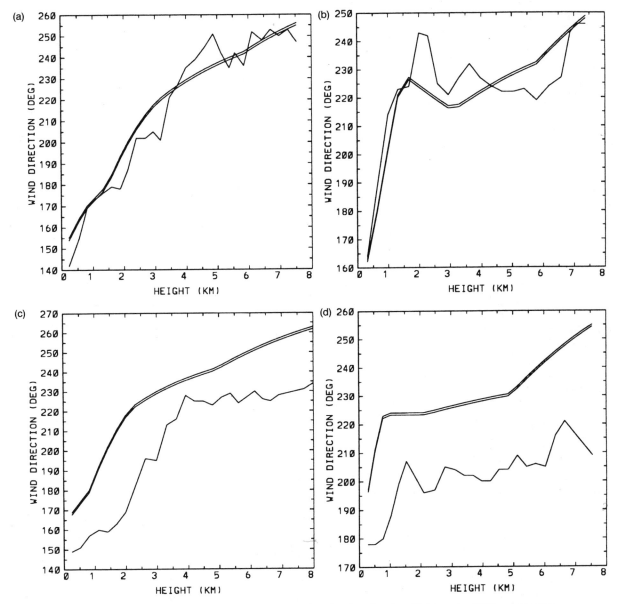

FIG. 11.27. Comparison of modeled and observed wind directions for phase II: (a) station WSC at 1800 PST, (b) station WSC at 0600 PST, (c) station WSF at 1800 PST, and (d) station WSF at 0600 PST.

APPENDIX

Description of RAMS

This model was developed at Colorado State University by Roger Pielke, Bill Cotton, Bob Walko, and Craig Tremback.

a. Equations

RAMS can be run in nonhydrostatic or hydrostatic form. The following treats the nonhydrostatic form.

RAMS quasi-Boussinesq equations of motion in tensor form are

$$\frac{\partial \rho_0 u_i}{\partial t} + \frac{\rho_0 \theta_0}{a} \frac{\partial ab^{ij}}{\partial x_j^*} \pi' = \mathrm{ADV}(\rho_0 u_i) + \rho_0 \mathrm{TURB}(u_i)$$

$$+ g\left(\frac{\theta'}{\theta_0} + 1.61 r_v - r_T\right)\delta_{13} + \epsilon_{ijk} f_3(\bar{u}_{ik} - u_{0k}).$$

The Exner function is defined as

$$\pi = c_p \left(\frac{P}{P_{00}}\right)^{R/c_p}$$

The temperature is related as

$$T = \theta \frac{\pi}{c_p} .$$

The time-dependent equation for the Exner function is given by Klemp and Wilhelmson (1978) and Durran (1981):

$$\frac{\partial \pi'}{\partial t} + \frac{1}{a} \frac{R}{c_v} \frac{\pi_0}{\rho_0 \theta_0} \frac{\partial (ab^{ij} \rho_0 \theta_0 u_i)}{\partial x_j^*} .$$

b. Dimensionality

Versions of RAMS have been developed for one, two, and three dimensions.

c. Grid

RAMS uses the staggered Arakawa C grid. Unrestricted grid spacing can be set from under 10 m to over 100 km. The horizontal coordinate is either Cartesian or polar stereographic. Horizontal spacing is assumed spatially constant in model coordinates but varies according to earth coordinates when polar stereographic coordinates are implemented. Vertical grid spacing is unrestricted, but typically starts at 10–100 m near the ground and tapers to 1 km at higher levels. RAMS can cover virtually any limited area up to a hemisphere.

A version of sigma height coordinates is used over complex terrain according to the methods of Gal-Chen and Sommerville (1975a,b), and derivatives are transformed according Clark (1977). To achieve high resolution in selected regions, nesting is performed according to the procedures of Clark and Farley (1984). In theory there is no limit to the number of nests that can be used, but in the current exercise three are used of 5-, 20-, and 80-km resolution.

The transformed coordinates take the form

$$x^* = x$$
$$y^* = y$$
$$z^* = H\left(\frac{z - z_s}{H - z_s}\right),$$

where z_s is the terrain height and H is the height of the model top. Derivatives therefore have the following form,

$$\frac{\partial A}{\partial x_i} = \frac{1}{a} \frac{\partial ab^{ij} A}{\partial x_j^*} ,$$

where a is given by

$$a(x^*, y^*) = 1 - \frac{z^s(x^*, y^*)}{H} = \frac{\partial z}{\partial z^*} ,$$

and b^{ij} is given by

$$b_{ij} = \begin{bmatrix} 1 & 0 & \frac{1}{a} \frac{\partial z}{\partial x^*}\left(\frac{z^*}{H} - 1\right) \\ 0 & 1 & \frac{1}{a} \frac{\partial z}{\partial y^*}\left(\frac{z^*}{H} - 1\right) \\ 0 & 0 & \frac{1}{a} \end{bmatrix} .$$

The Cartesian velocities can then be described as

$$u^* = u$$
$$v^* = v$$
$$w^* = (uab^{13} + vab^{23} + w)a^{-1}.$$

d. Analysis methodology

Atmospheric data are first collected from sources such as gridded pressure level datasets, standard and special soundings, and surface stations. All observed atmospheric data are first interpolated to isentropic surfaces. These data are then combined using a Barnes objective analysis. The analyzed data are then interpolated to the model grid.

e. Model initialization procedure

Usually RAMS is initialized from 3D atmospheric fields that are interpolated from analyzed isentropic data. For a small domain a single sounding is used. No special initialization procedure is used, although boundary nudging can be used when time series data are available.

f. Solution—Numerical technique

Initial value prognostic equations are solved using finite differences. This includes a prognostic Exner equation. For this equation a time split method is used to handle the fast propagation of acoustic waves. Leapfrog or forward time differencing is used for the time integration. Acoustic modes are solved using the Crank–Nicholson method, but most other integration is explicit. Advective schemes are second-, fourth-, or sixth-order accurate. Some horizontal diffusion or fourth-order filtering is necessary to damp small wiggles.

g. Boundary treatment

At the upper boundary either a rigid lid, where vertical velocities and gradients are set to zero, or a semi-open boundary can be used to let gravity waves propagate out from the domain. When the rigid lid is used, Rayleigh damping is used to nudge solutions or damp gravity waves. Lateral boundaries use radiative conditions on the normal velocity component. A variety of other conditions, including nudging in inhomogeneous conditions, can also be used.

h. Canopy treatment

RAMS contains a canopy parameterization including the effects of vegetation on the soil, radiation, sensible heat exchange, and evapotranspiration.

i. Soil treatment

RAMS contains a user specified number of levels of soil. The top layer of the soil exchanges heat, moisture, and radiation with the atmosphere. Twelve standard soils are cataloged for use in the model.

j. Surface boundary treatment

The model uses the surface-layer parameterization of Louis (1979) for treatment of vertical heat, vapor, and momentum fluxes. The height of the surface layer is taken as the first integrated θ point above the surface.

k. Special conditions

RAMS has been tested in many conditions including sea breezes, and many synoptic patterns.

l. Cumulus parameterization

RAMS optionally uses a modified Kuo or Fritsch–Chappel cumulus parameterization method.

m. Radiation

Two longwave and shortwave radiation schemes are available in RAMS. The more complex accounts for effects of clouds, and both account for water vapor, carbon dioxide, and ground and vegetation effects.

n. Stable precipitation

RAMS contains microphysical bulk parameterizations for stable and convective precipitation. Cloud water, rain, snow, and other species and microprocesses are covered.

o. Coding practices

RAMS has been coded in Fortran 77 and Fortran 90. RAMS is documented, and journal and report references are available. RAMS runs in Unix and on a variety of computers including CRAYs, Stardent, and IBM RISC, and on PCs.

Chapter 12

HOTMAC: Model Performance Evaluation by Using Project WIND Phase I and II Data

TETSUJI YAMADA

Yamada Science and Art Corporation, Los Alamos, New Mexico

TEIZI HENMI

United States Army Atmospheric Science Laboratory, White Sands Missile Range, New Mexico

12.1. Introduction

A three-dimensional mesoscale model, HOTMAC (Higher Order Turbulence Model for Atmospheric Circulation), was applied to simulate the data collected during Project WIND. Project WIND collected comprehensive meteorological data over complex terrain and variable land use on a scale ranging from 200 km × 200 km to 5 km × 5 km in and around the Sacramento River valley of northern California (Cionco 1991). (See also chapter 8.)

HOTMAC simulated data in two cases: summer (phase I) and winter (phase II) conditions. The data consisted of measurements from 25 surface sensors and 5 upper-air sounding stations within an 80 km × 80 km domain (Fig. 12.1). Surface data included 1-min averages of wind speed, wind direction, temperature, relative humidity, and pressure, as well as precipitation measurements.

The phase I was characterized by typical summertime anticyclonic conditions over Sacramento Valley. Southerly flows (upslope winds) during the day and northerly flows (downslope winds) during the night were the characteristic flow patterns. Under severe weather conditions (phase II), the dominant forces are large-scale pressure gradients, and diurnal and spatial variations due to local forcing become insignificant. For successful simulations of severe weather phenomena, a mesoscale model must be able to incorporate the large-scale weather variations. A four-dimensional data assimilation method (Anthes 1974; Hoke and Anthes 1976) was found to be a simple but effective way to incorporate the large-scale forcing into a mesoscale model.

Section 2 reviews model equations, initial and boundary conditions, a four-dimensional data assimilation method, and finite-difference equations for numerical solutions. Section 3 explains model performance evaluation methods. Sections 4 and 5 discuss,

respectively, phase I and phase II data simulations and statistical evaluations of the model performance. A summary is given in section 6.

12.2. Model description

a. Model equations

The model used here is referred to as HOTMAC. HOTMAC is based on the concept of the ensemble average: we seek the solutions in terms of mean values and deviations from the mean (turbulence). An ensemble average model is a realistic choice, in terms of accuracy and computational time, for three-dimensional atmospheric simulations.

The disadvantage of an ensemble model is the fact that the number of unknowns exceeds the number of equations. Thus, it is necessary to introduce closure hypotheses that relate unknown higher-order moments to the known lower-order moments. Our closure hypotheses are based on those proposed by Mellor (1973). A hierarchy of turbulence closure models was proposed where complex model equations were systematically simplified (Mellor and Yamada 1974). Users can select a different level model to balance between the desired accuracy and computer time constraint. A word "level" is used to differentiate the completeness of model physics: the level 4 model is the most complete and the level 1 model is the simplest. The level 2.5 model (Mellor and Yamada 1982) was found to be a good choice for three-dimensional simulations in terms of the accuracy and computation time. A summary of the model applications is given in a review paper by Mellor and Yamada (1982).

The basic equations of HOTMAC were described in detail by Yamada and Bunker (1988, 1989) and thus only a brief review is given here, together with a summary of the equations in the appendix.

The governing equations are conservation equations for mass, momentum, potential temperature devia-

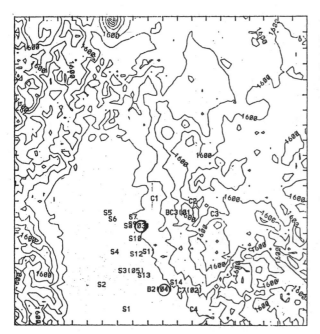

FIG. 12.1. The model domain (200 km × 200 km) where terrain is contoured by solid lines with an increment of 400 m. The lowest contour is 400 m above the mean sea level. The locations of surface stations are indicated by the characters. The numbers in the parentheses represent upper-air stations.

tions, water vapor, and turbulence kinetic energy. It is noted that the deviations of potential temperature from the large-scale mean values were solved instead of the absolute values of potential temperature. This modification was found to be useful to reduce computational errors and to keep predicted wind fields realistic (Yamada and Bunker 1989). The magnitude of the potential temperature is about 300 K, but the deviations from the large-scale values are only 10 K or less. Thus, one or two more significant figures can be carried throughout the computations if temperature deviations, instead of absolute temperatures, are used.

Another source of errors was associated with a terrain-following vertical coordinate used in the model. A terrain-following coordinate is useful to simplify the treatment of the surface boundary conditions. On the other hand, numerical diffusion inherent to the finite-difference numerical scheme resulted in spurious pressure gradients, particularly when temperature gradients were large in the vertical direction. Potential temperature deviations from the large-scale values varied very little in the horizontal directions in the terrain-following coordinate system. Thus, numerical diffusion becomes small because diffusion is proportional to the higher derivatives of potential temperature deviations in the horizontal directions. The large-scale temperature varied with height, but it was assumed to be constant with time and in the horizontal directions.

The present model assumes hydrostatic equilibrium and uses the Boussinesq approximation. Therefore, in

theory, the model is valid for flows where the vertical scale of the modeled system is small compared with its horizontal scale (hydrostatic equilibrium), and for small temperature variations in the horizontal directions (Boussinesq approximation). Both assumptions are easily satisfied with a horizontal grid spacing of 5 km used in this study.

b. Data assimilation method

Mesoscale models such as HOTMAC are able to forecast wind distributions associated with the pressure gradients generated in the computational domain. However, any variations produced outside the computational domain must be incorporated into mesoscale models through additional forcing terms.

Variations of large-scale wind distributions were incorporated into the equations of motion through a technique referred to as "nudging" or "Newtonian relaxation" methods (Anthes 1974; Hoke and Anthes 1976). The terms $C_n(U_t - U)$ and $C_n(V_t - V)$ were added to the equations of motion for the east–west and north–south components, respectively. Here, U_t and V_t are "target" wind components for the corresponding wind components U and V, respectively, and were computed, as in (12.1) and (12.2), from observed winds and geostrophic winds. Equations (12.1) and (12.2) were derived from the equations of motion where horizontal homogeneity was assumed (Yamada and Bunker 1989):

$$U_t = U_{\text{obs}} - \frac{f}{G}(V_{\text{obs}} - V_g), \qquad (12.1)$$

$$V_t = V_{\text{obs}} + \frac{f}{G}(U_{\text{obs}} - U_g). \qquad (12.2)$$

Here, U_{obs} and V_{obs} are observed wind components, and U_g and V_g are geostrophic wind components. Thus, U_t and V_t are, in general, different from the observed wind components. The nudging method is a simple but effective way to incorporate large-scale wind variations into a mesoscale model.

In (12.1) and (12.2), U_{obs} and V_{obs} were determined from upper-air soundings in the following manner. First, linear interpolation of horizontal wind vector components was performed as a function of altitude on the upper-air sounding data. Data from station 03 were used for phase I, and data from station 04 were used for phase II. The locations of stations 03 and 04 are shown circled in Fig. 12.1. Then, hourly wind profiles between the sounding periods were obtained by linear interpolation. The so-called observed winds at other grid points were obtained by multiplying the winds at station 03 or 04 by a factor $(z - z_g)/(H - z_g)$ in order to satisfy, approximately, the mass conservation requirement. Here, z is the altitude, z_g the terrain height, and H the height of the top of model atmosphere. Wind data were read hourly at 1 h before the simulation started. For example, winds at 1200 PST were read in at 1100 PST and nudged from 1100 to 1200 PST. This process was repeated for the entire simulation period.

There were five upper-air measurement stations located relatively close to each other within the 80 km × 80 km study area. Thus, the wind data at any one of these stations should represent well the large-scale wind profile for the computational domain. But this was found to be not quite true: there were variations among soundings. An alternative approach would be to interpolate or extrapolate the soundings at five stations and produce observed winds at every grid point. Since the sounding stations were located relatively close to each other in the central area, the extrapolation to the entire computational domain would have introduced significant errors in the observed wind speeds toward which modeled winds would have been nudged.

We have chosen to use the wind data at one sounding station mainly for simplicity. Thus, U_{obs} and V_{obs} in (12.1) and (12.2) were constant in the horizontal directions. The assumption of the horizontal homogeneity for large-scale winds in the 200 km × 200 km computational domain is probably valid under fair weather conditions such as for phase I case. However, the assumption fails when severe mesoscale disturbances, such as fronts, prevail. Then, it is important to incorporate spatial variations of U_{obs} and V_{obs}. Unfortunately, no upper-air soundings were taken during the frontal passage for the phase II case. Thus, the variation of wind with time and space could not be incorporated in the model. As seen in section 12.4, standard deviations of the observed surface winds were large during the frontal passage. On the other hand, the modeled surface winds show small standard deviations because the large-scale wind variations could not be incorporated into HOTMAC. More discussions on this matter are given in section 12.5.

It is noted that only horizontal wind components were nudged toward observations. Temperature, mixing ratio of water vapor, and other prognostic variables were initialized by using soundings at the beginning of simulations, but they were not assimilated.

c. Boundary conditions

Surface boundary conditions were constructed from the empirical formulas of Dyer and Hicks (1970) for nondimensional wind and temperature profiles. The temperatures in the soil layer were obtained by solutions of the heat conduction equation. The appropriate boundary conditions were the heat balance at the soil surface and specification of the soil temperature at a certain depth. The lateral boundary values were obtained by integration of the corresponding governing equations except that the variations in the horizontal directions were neglected.

Effects of tall tree canopies on wind and temperature distributions were studied (Yamada 1982). Tall trees

play significant roles in determining wind and temperature distributions in the surface layer. For example, wind speeds reduce considerably in the forest canopy due mainly to the surface drag induced by leaves, stems, and branches (e.g., Oliver 1971). Tall trees also modify temperatures within and above the canopy. For example, during the day initial warming of the upper canopy results in an unstable layer above the canopy, and a stable layer within the canopy (e.g., Hosker et al. 1974). This stability structure is reversed during the night because the rate of cooling due to longwave radiation is greatest in the upper canopy.

Following Wilson and Shaw (1977), we assumed that pressure forces contributed the major portion of the total due to the canopy. The form drags appeared in the equations of motion were modeled to be proportional to the plant area density, wind speed, and wind components. The proportionality constant was referred to as a drag coefficient. The parameterization ensures that the direction of the drag force is always opposite to the wind direction. Thus, wind speed will decrease.

d. Numerical method

The governing equations were integrated by use of the alternating direction implicit method (Richtmyer and Morton 1967). A time increment was chosen to be 90% of the value of Δ/U, where Δ is a grid spacing and U is the maximum value of the velocity components or the wave propagation speed. The propagation speeds of gravity waves in the atmosphere were computed based on the shallow-water theory and the density stratification of large-scale temperature profile.

12.3. Model performance evaluation method

For the present study, surface data of wind direction, wind speed, temperature, and dewpoint temperature were available throughout the 24-h period at 21 surface stations for both the phase I and II measurement periods. Upper-air sounding data were available every 2 h at five sites with some exceptions in the phase II case.

For surface data, mean values; standard deviations; systematic and unsystematic components of the root-mean-square differences, as well as the total root-mean-square differences; and agreement measures were calculated hourly for wind, temperature, and dewpoint. No statistical comparison was made for upper-air data, due to insufficient number of data.

- Mean

$$\bar{x} = \frac{1}{N} \sum_k x(k), \qquad (12.3)$$

where $x(k)$ is one of the meteorological variables at kth station and N is the total number of stations. Mean values for both simulations and observations were calculated.

- Standard deviation

$$\sigma = \left\{ \sum_k \frac{[x(k) - \bar{x}]^2}{N} \right\}^{1/2}. \qquad (12.4)$$

- Root-mean-square differences

Use of systematic and unsystematic components of the root-mean-square differences as well as the total root-mean-square difference (rmsd_s, rmsd_u, and rmsd, respectively) was recommended by Willmott et al. (1985) for quantitative evaluation of model performance. Steyn and McKendry (1988) used these parameters to evaluate the performance of the Colorado State University's mesoscale model (Mahrer and Pielke 1977b, 1978) for a sea-breeze circulation in complex terrain. These parameters were also used by Ulrickson and Mass (1990) for the evaluation of the same model used for simulation of mesoscale circulation over the Los Angeles basin.

These parameters are defined as

$$\mathrm{rmsd}_s = \left\{ \frac{1}{N} \sum_k [x_m^*(k) - x_o(k)]^2 \right\}^{1/2}, \quad (12.5)$$

$$\mathrm{rmsd}_u = \left\{ \frac{1}{N} \sum_k [x_m^*(k) - x_m(k)]^2 \right\}^{1/2}, \quad (12.6)$$

$$\mathrm{rmsd} = \left\{ \frac{1}{N} \sum_k [x_m(k) - x_o(k)]^2 \right\}^{1/2}, \quad (12.7)$$

where N is the number of stations (evaluation points), x_m and x_o are modeled and observed values, respectively, and $x_m^* = a + bx_o$, where a and b are the parameters associated with an ordinary least-squares linear regression between x_o and x_m (Steyn and McKendry 1988).

The rmsd_s is an estimate not only of a model's offset bias but also of any linear variation in the model's bias. The rmsd_u is a measure of the nonlinear discrepancy between simulations and observations (Ulrickson and Mass 1990). The definitions of systematic and unsystematic rmsd apply to the spatial coherence of each hourly set of simulations and observations and do not address the errors at individual sites.

- Agreement measure

$$A = 1 - \frac{\sum\limits_k [x_m(k) - x_o(k)]^2}{\sum\limits_k [|x_m(k) - \overline{x_m}| + |x_o(k) - \overline{x_o}|]^2}. $$

$$(12.8)$$

This dimensionless index has a theoretical range of 1.0 (for perfect agreement) to 0.0 (for no agreement). This parameter was also used by Steyn and McKendry (1988) and Ulrickson and Mass (1990) for their studies of model evaluation.

12.4. Phase I data simulation

Phase I data were collected in June and July 1985. The Mesomet Advisory Panel has selected the data taken on Julian days 178 and 179 (27 and 28 June) of 1985 to be used for simulations and verification.

Airflows on day 178 were characterized as typical for a summer day: southerly (upslope) flows during the daytime and northerly (downslope) flows during the nighttime. The southerly flow was consistent with weak marine incursion resulting from the winds, which initiated from the San Francisco Bay area, moved inland, and diverted to the north and south by the Sierra Nevada along the Sacramento Valley.

a. Initial values

An initial wind profile at station 03 was first constructed by assuming a logarithmic variation with u_* = 0.2 m s^{-1} and z_0 = 0.1 m from the ground up to the height where the wind speed reaches an ambient value, which was 5 m s^{-1} in this case. Initial wind profiles at other grid locations were obtained by multiplying the winds at station 03 by a factor $(z - z_g)/(H - z_g)$ to satisfy, approximately, mass conservation. The wind direction was initially assumed to be from the southsouthwest (210°) throughout the atmosphere. The vertical profile of potential temperature was initialized based on an upper-air sounding data taken at station 03 at 0900 PST of day 178. Initial potential temperatures were assumed to be uniform in the horizontal directions. Initial values of water vapor were also based on observations. The turbulence kinetic energy and the length scale were initialized by using the initial wind and temperature profiles, and the relationships determined by the level 2 turbulence-closure model (Yamada 1975). The level 2 model assumes a balance between the production and dissipation in the turbulence kinetic energy equation.

The calculation started at 0900 PST of day 178 and lasted until 0900 of day 179. However, model calculation was set four hours before the start of simulation to allow initial adjustment. At 1 h before the start of simulation, observed wind data were read in and model winds were nudged for 1 h toward observations, as described in the previous section. Temperature, water vapor, and other prognostic variables were not nudged toward observations.

b. Surface data

1) DIURNAL PATTERN

The evolutions of wind direction and wind speed at 10 m above the ground, temperatures, and dewpoint temperature at 2 m above the ground are given in Fig. 12.2 at four selected stations: S2, S10, S14, and C3. In Fig. 12.2, hourly values of observations are plotted with lines with solid circles, and simulations are plotted with solid lines.

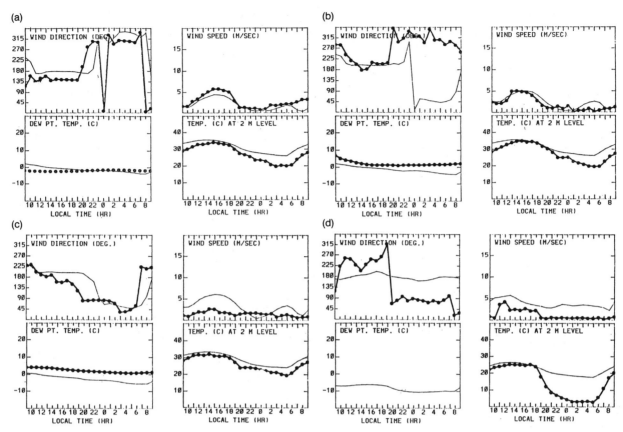

FIG. 12.2. Time evolutions of the observed (dots) and simulated (thin line) surface meteorological variables (phase I):
(a) station S2, (b) station S10, (c) station S14, and (d) station C3.

Station S2 was located in the valley. Wind direction shift from the south during the daytime to the north during the nighttime was well simulated. The diurnal variations of the modeled wind speeds are also in good agreement with the observations.

The modeled and observed wind directions showed disagreement after sunset at station S10, which was located near the foothills of the Sierra Nevada: the modeled wind directions were northeasterly, but the observations were predominantly northwesterly. It is noted that the observed wind speeds during the nocturnal period were approximately 1 m s^{-1}. Wind directions for such small winds cannot be determined accurately and are subject to large errors.

At station S14, which was located at the western slope of the mountain range, the observed wind directions changed gradually from the southwest to the east, but the model produced a rather sudden shift of wind directions from the south to the east after the sunset. The modeled and observed wind speeds did not agree well during the daytime.

At station C3, the observed wind directions changed from the southwest in the day to the east in the night, but the simulated wind directions remained in the south throughout the 24-h period. The modeled and observed wind speeds did not agree well either. At this

station, observed wind speeds remained very small during the nighttime. The model failed to simulate small wind speeds.

In general the modeled temperatures were in good agreement with observations during the daytime. But during the nighttime the differences between the simulated and observed temperatures became large. Particularly at C3, where the model failed to simulate a large temperature drop at night. Station C3 was located in a meadow that was surrounded by high mountains, indicating that the nighttime temperature drop could be associated with cold-air drainage from the surrounding mountains. The model did not reproduce the cold-air drainage flows that were considered to be subgrid-scale phenomena with a 5-km grid spacing used in the model.

In general, the modeled temperatures during the nocturnal periods were much higher than the observations. Throughout the 24-h period, the dewpoint temperatures stayed low as seen from Fig. 12.2. No observation was available at station C3.

2) STATISTICAL EVALUATION

Using the surface data measured at 21 stations, mean wind directions, wind speeds, and standard deviations

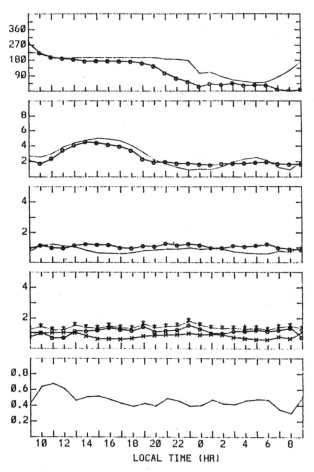

FIG. 12.3. Statistical parameters for winds at 10 m above ground (phase I). From top, mean wind direction, mean wind speed (m s^{-1}), standard deviation of wind speed, root-mean-square differences (rmsd—asterisks, rmsd$_u$—crosses, and rmsd$_s$—dots), and agreement measure of wind speed. Dots indicate observed values, and the thin curve indicates simulated values.

of wind speed were calculated hourly for the simulations (solid line) and observations (circles) as shown in Fig. 12.3. Also shown are hourly rmsd, rmsd$_s$, and rmsd$_u$, and agreement measures of wind speeds. Mean wind directions were calculated from the means of horizontal wind components.

From the top two panels of the figure, it is clear that the overall patterns of the observed and simulated surface wind fields agreed well except for several hours after sunset.

Mean wind directions of the observed winds were from the southwest during the day, shifted gradually to the north after sunset, and then stayed to the northeast during the nighttime. The simulated mean wind directions agreed well with observations during the daytime, but the shift to the north was delayed for several hours, and after midnight, the wind stayed in the north. The modeled mean wind speeds were in good agreement with observations throughout the 24-h period.

The modeled standard deviations of wind speed were comparable with those observed throughout the simulation period. The rmsd, rmsd$_s$, and rmsd$_u$ show no diurnal variations, and both rmsd$_s$ and rmsd$_u$ are of roughly in the same magnitude. The unsystematic component, rmsd$_u$, represents the irreducible deviation between observed and simulated results, while systematic component represents trendlike differences between the observed and simulated distributions (Steyn and McKendry 1988).

The highest value (0.65) for the agreement measure was obtained at 1100 PST of day 178. But, after 1200 PST, the agreement measures remained approximately at 0.5 without any systematic trend. The absence of a trend in the agreement measures indicates that the model was capable of simulating the diurnal wind variations over the Project WIND area.

The mean values, standard deviations, rmsd's, and agreement measures for the temperatures at 2 m above the ground were also calculated hourly, and the results are shown in Fig. 12.4. There were good agreements between the simulations and observations during the daytime, but during the nighttime, noticeable differences developed between the observed and simulated temperatures. The modeled surface temperatures were much higher than observations during the nighttime, resulting in large values for rmsd$_u$ and rmsd$_s$, and in small values for the agreement measures.

c. Upper-air data

Vertical distributions of wind direction and wind speed, temperature, and dewpoint temperature at 0900, 1500, and 2100 PST of day 178, and 0300 PST of day 179 at station 1 are shown in Fig. 12.5. In Fig. 12.5, simulations are shown by the solid lines and observations by the circles. Good agreement was obtained between modeled and observed vertical profiles of area-mean wind speed, wind direction, temperature, and dewpoint throughout the entire simulation period. It is noted that no nudging was applied in the temperature and mixing ratio equations.

12.5. Phase II data simulation

Phase II observations were carried out in January and February 1986 to study wintertime cyclonic activity (see chapter 8 and Cionco 1991). The Mesomet Advisory Panel has selected Julian days 32 and 33 (February 1 and 2) of 1986 to be used for simulations and verification.

During days 32 and 33, the area was under the influence of a low pressure cell. The period was cold and wet starting with partly cloudy skies increasing to heavy overcast and the onset of rain and rain showers throughout the nighttime with moderate to strong gusty winds. Winds over the project area were predominantly southerly throughout the study period. Observation

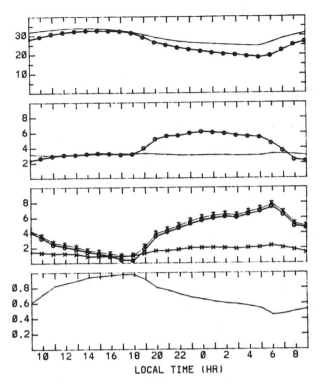

FIG. 12.4. Statistical parameters for temperatures at 2 m above ground (phase I). From top, mean temperature (°C), standard deviations, root-mean-square differences (rmsd—asterisks, rmsd$_u$—crosses, and rmsd$_s$—dots), and agreement measure. Dots indicate observed values, and the thin curve indicates simulated values.

started at 1000 PST of day 32 and ended at 1000 PST the next day. Surface data similar to those for the phase I were available. But, due to strong winds in the frontal zone, several upper-air sounding data were not available during the early morning periods of day 33.

a. Initial values

The initialization method for the phase II data simulation was similar to that for the phase I data simulation, except that upper-air sounding data taken at station 04, instead of station 03, were used.

b. Surface data

1) DIURNAL PATTERN

Evolutions of wind direction and wind speed at 10 m above the ground, and temperature and dewpoint temperature at 2 m above the ground for stations S1, S5, S14, C3, and C7 are plotted in Fig. 12.6.

It is seen clearly that temperatures and dewpoint temperatures varied very little during the 24-h period, and the agreements between the observations (lines with solid circles) and the simulations (solid lines) were excellent. Throughout the simulation periods, the model produced steady wind flows with wind directions

ranging from the southwest to the south and wind speeds of about 5–7 m s^{-1} at all the stations except at C3. At station C3, simulated wind speeds reached 5 m s^{-1} in the early morning of day 33, whereas the observed wind speeds were less than 2 m s^{-1} during the entire simulation period.

The observed surface wind directions remained in the southwest to the west, except at station C3. Generally, wind speeds increased gradually in the evening of day 32 and peaked in the early morning hours of day 33, corresponding to the frontal passage.

However, the observed wind directions and wind speeds at station C3 showed quite different evolutions; wind directions varied widely and wind speeds remained small throughout the 24-h period. Station C3 was located in a meadow surrounded by high mountains. Thus, station C3 might be protected from the influences of the frontal passage. The observed strong cooling in the morning of day 179 (Fig. 12.2d) and calm wind speeds at C3 (Fig. 12.6d) are the examples of subgrid-scale phenomena that the model could not simulate with a coarse grid spacing of 5 km.

The distance between the stations S14 and C7 was only about 7 km, and the altitudes of both stations were similar, but the evolutions of wind speed were significantly different. At station C7 (Fig. 12.6e), notable peaks of wind speed greater than 10 m s^{-1} were recorded in the early morning hours of day 33, but at station S14 (Fig. 12.6c) wind speeds increased only slightly during the corresponding hours. A close examination of the terrain data revealed that station S14 was well protected to the southerly flows compared to station C7. This is another example of subgrid-scale phenomena that could not be simulated by the model due to the coarse grid spacing used in the simulations.

As seen from Figs. 12.6a–e, the model did not produce wind speed maxima associated with frontal passage. As noted earlier, upper-air soundings in the early morning hours (0400, 0600, and 0800 PST) of day 33 at all stations were missing. Therefore, for these periods, winds interpolated from 0200 and 1000 PST soundings at station 04 were used for nudging, which were significantly smaller than the winds associated with the frontal passage.

2) STATISTICAL EVALUATIONS

As for the phase I case, the evolutions of the mean wind directions, wind speeds, standard deviations, rmsd's, and agreement measures were calculated (Fig. 12.7). It is clear that the modeled wind directions are in good agreement with observations throughout the entire period; wind directions remained in the southeast both for observations and simulations. Noticeable differences between the modeled and observed wind speeds occurred in the early morning of day 33 when a front passed the area. The observed mean wind speeds became large, but the model failed to produce peak

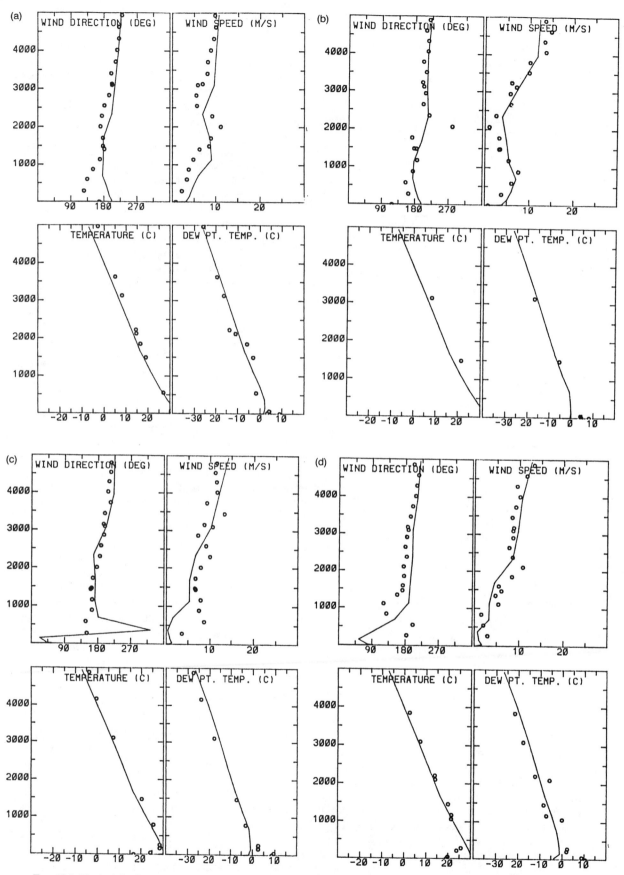

FIG. 12.5. Vertical distributions of wind direction, wind speed, temperature, and dewpoint temperature at station 04 (phase I): (a) 1100 PST day 178, (b) 1700 PST day 178, (c) 0100 PST day 179, and (d) 0700 PST day 179.

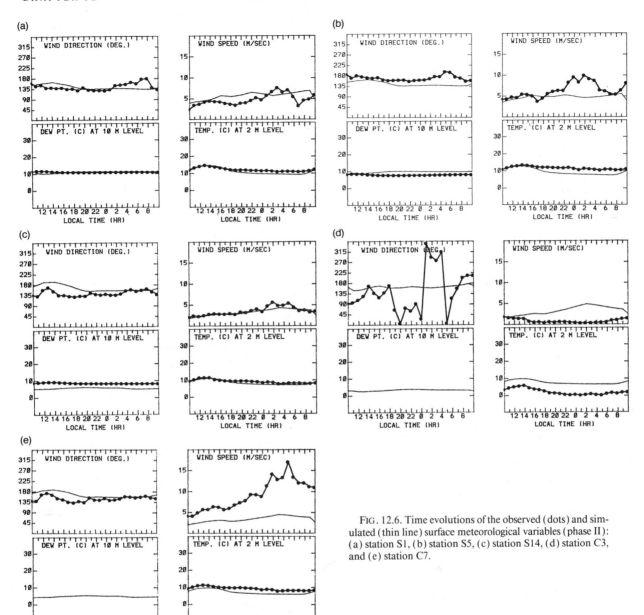

FIG. 12.6. Time evolutions of the observed (dots) and simulated (thin line) surface meteorological variables (phase II): (a) station S1, (b) station S5, (c) station S14, (d) station C3, and (e) station C7.

wind speeds in the early morning hours of day 33. The wind speed discrepancies occurred because the modeled winds were nudged toward winds that were obtained by interpolating the soundings at 0200 (before the frontal passage) and 1000 PST (after the frontal passage). No soundings were taken during the frontal passage.

Standard deviations of the observed wind speeds were much greater than those simulated throughout the 24-h period. The discrepancies were caused by the lack in the model of forcing due to the frontal passage.

The rmsd and rmsd$_s$ became large during the period of the frontal passage, indicating large discrepancies in the modeled and observed wind distributions.

The evolutions of statistical parameters for temperatures at 2 m above the ground are shown in Fig. 12.8. Throughout the study period, observed temperatures did not vary significantly and were in good agreement with the simulated temperatures. Standard deviations of the observed temperatures and rmsd's were considerably smaller in the phase II case than in the phase I case indicating that the area was under the strong influence of large-scale weather conditions.

The agreement measures were approximately 0.6 throughout the 24-h period.

Maxima of the standard deviations for the observed temperatures and rmsd's, and a minimum of the agreement measures in the early morning (between

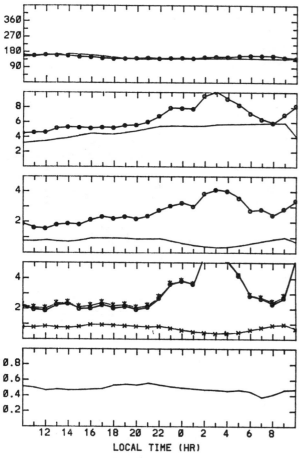

FIG. 12.7. Statistical parameters for winds at 10 m above ground (phase II). From top, mean wind direction, mean wind speed, standard deviation of wind speed, root-mean-square differences (rmsd—asterisks, rmsd$_u$—crosses, and rmsd$_s$—dots), and agreement measure of wind speed. Dots indicate observed values, and the thin curve indicates simulated values.

0200 and 0500 PST) were due to errors in the observed temperatures at some stations.

c. Upper-air data

Vertical profiles of the wind direction, wind speed, temperature, and dewpoint at different hours for station 05 are shown in Fig. 12.9. Note that upper-air winds taken at station 04 were used for assimilations. No nudging was applied in the temperature and mixing ratio equations.

In general, the simulated variables are in good agreement with the observations until 0000 PST of day 33. The model successfully simulated the increase of wind speeds with time and strong wind shears below 1000-m levels (see the vertical profiles at 0000 PST of day 33). The simulated wind speeds at 1000 PST of day 33 were lower than the observations.

12.6. Summary

A three-dimensional mesoscale model, HOTMAC, was applied to simulate the Project WIND phases I

and II data, which were taken under quite different meteorological conditions. The phase I (summer) was characterized by typical summertime anticyclonic conditions over Sacramento Valley in northern California. Southerly flows (upslope winds) during the day and northerly flows (downslope winds) during the night were the characteristic flow patterns. A four-dimensional data assimilation method resulted in good agreements between the observed and simulated meteorological variables.

Statistical parameters such as mean values and standard deviations of both observed and simulated variables; systematic and unsystematic components of the root-mean square differences, as well as the total root-mean-square difference (rmsd$_s$, rmsd$_u$, and rmsd, respectively); and agreement measures were calculated hourly by using the surface station data. It was shown that these parameters were useful for evaluations of model performance.

Evaluations of the statistical parameters indicated HOTMAC's capabilities in simulating successfully the meteorological data taken both under fair (phase I) and severe (phase II) weather conditions. Under fair weather conditions (phase I), the dominant forces that control wind flows are the pressure gradients resulting from the differential heating and cooling over the in-

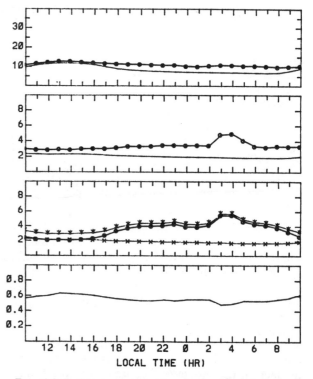

FIG. 12.8. Statistical parameters for temperatures at 2 m above the ground (phase II). From top, mean temperature, standard deviations, root-mean-square differences (rmsd—asterisks, rmsd$_u$—crosses, and rmsd$_s$—dots), and agreement measure.

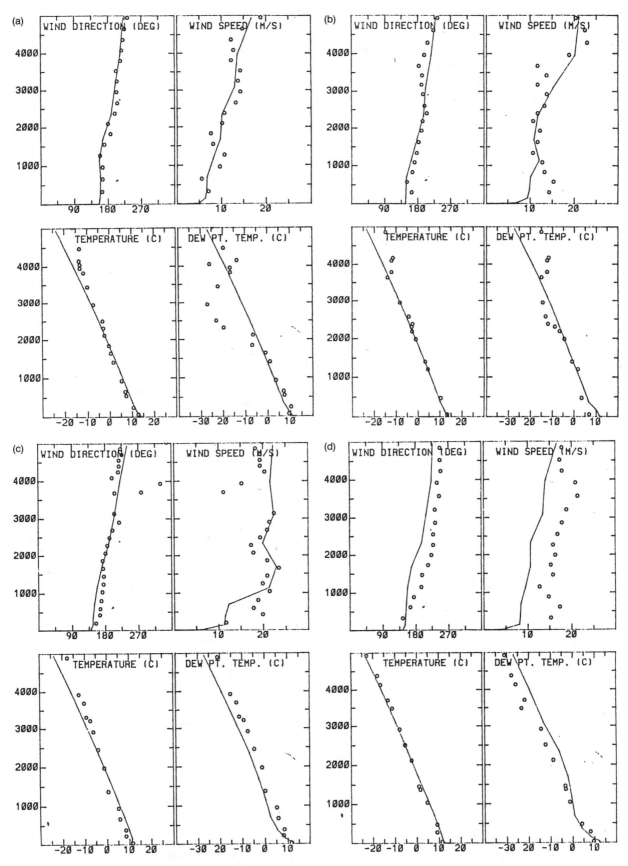

FIG. 12.9. Vertical distributions of wind direction, wind speed, temperature, and dewpoint temperature at station 05 (phase II):
(a) 1200 PST day 32, (b) 1600 PST day 32, (c) 0000 PST day 33, and (d) 1000 PST day 33.

homogeneous surfaces (terrain, land–water distribution, soil characteristics, vegetation distribution, etc.).

HOTMAC reproduced well the observed diurnal and spatial variations of meteorological variables (wind speed, wind direction, temperature, mixing ratio of water vapor) at most stations. But the model failed to simulate large temperature drop observed during the nighttime at station C3, which was surrounded by high mountains. To simulate such subgrid-scale phenomena, a horizontal grid spacing much smaller than the present 5-km spacing would be required. Other factors that significantly influence the surface temperature variations are soil moisture, soil characteristics, and vegetation distributions. Very little information on these parameters was available for the simulations.

Under severe weather conditions (phase II), the dominant forces are large-scale pressure gradients and diurnal and spatial variations due to local forcing become insignificant. For successful simulations of severe weather phenomena, a mesoscale model must be able to incorporate the large-scale weather variations. A nudging method was found to be a simple but effective way to incorporate large-scale forcing into a mesoscale model. It is important that large-scale data used for nudging are accurate. Unfortunately, large-scale data may not always be available, as was the case for the phase II simulations. Ideally, both temporal and spatial variations of the large-scale weather conditions should be used for nudging in mesoscale models. With such data, model performance measured by rmsd would have improved significantly.

Acknowledgments. This work was partially supported by U.S. Army Atmospheric Sciences Laboratory, White Sands Missile Range, New Mexico, U.S.A.

APPENDIX

Description of HOTMAC

This model has been developed by T. Yamada for applications in mesoscale meteorology.

a. Equations

The model equations (see Yamdada and Bunker 1989) were first developed with the hydrostatic treatment of pressure but recently have also been formulated in a nonhydrostatic form. Key equations for the hydrostatic Boussinesq version are as follows:

$$\frac{DU}{Dt} = f(V - V_g) + g\frac{\bar{H} - z^*}{\bar{H}}\left(1 - \frac{\langle\theta_v\rangle}{\theta_v}\right)\frac{\partial z_g}{\partial x}\frac{\partial}{\partial x}\left(K_x\frac{\partial U}{\partial x}\right)$$

$$+ \frac{\partial}{\partial y}\left(K_{xy}\frac{\partial U}{\partial y}\right) + \frac{\bar{H}}{H - z_g}\left[\frac{\partial}{\partial z^*}(-\overline{uw})\right] - \eta C_d a(z)US + G(U_o - U),$$

$$\frac{DV}{Dt} = -f(U - U_g) + g\frac{\bar{H} - z^*}{\bar{H}}\left(1 - \frac{\langle\theta_v\rangle}{\theta_v}\right)\frac{\partial z_g}{\partial y}\frac{\partial}{\partial x}\left(K_{xy}\frac{\partial V}{\partial x}\right)$$

$$+ \frac{\partial}{\partial y}\left(K_y\frac{\partial V}{\partial y}\right) + \frac{\bar{H}}{H - z_g}\left[\frac{\partial}{\partial z^*}(-\overline{vw})\right] - \eta C_d a(z)VS + G(V_o - V),$$

where

$$W^* = \frac{\bar{H}}{H - z_g}W + \frac{z^* - H}{H - z_g}\left(U\frac{\partial z_g}{\partial x} + V\frac{\partial z_g}{\partial y}\right)$$

and

$$\frac{D}{Dt} = \frac{\partial}{\partial t} + U\frac{\partial}{\partial x} + V\frac{\partial}{\partial y} + W^*\frac{\partial}{\partial z^*}.$$

The geostrophic winds are obtained using the geostrophic assumption

$$fU_g = fU_g(\bar{H})\frac{\langle\theta_v\rangle}{\langle\theta_v(H)\rangle}$$

$$+ g\frac{H - z_g}{\bar{H}}\int_{z^*}^{\bar{H}}\frac{1}{\langle\theta_v\rangle}\frac{\partial}{\partial y}\Delta\theta_v dz'$$

$$- \frac{g}{\bar{H}}\frac{\partial z_g}{\partial y}\int_{z^*}^{\bar{H}}\frac{1}{\langle\theta_v\rangle}\Delta\theta_v dz'$$

and similarly for fV_g.

Equations are formulated for the turbulent kinetic energy $q^2/2$, q^2l, the potential temperature deviation, and the mixing ratio of water vapor. Turbulent closures are from simplified versions of the second moment closures.

b. Dimensionality

One-, two-, and three-dimensional versions have been formulated.

c. Grid and coordinates

A terrain-following vertical coordinate system is used.

In three dimensions, a typical grid is $40 \times 40 \times 16$, and a $25 \times 25 \times 16$ nested region has also been accommodated, though other configurations can be used.

Horizontal grid spacing from a few hundred meters to over 20 km have been implemented. The model runs on a variety of domains. Some of the domains used are given in Table 12A.1.

d. Model initialization procedure

An initialization procedure has been developed to satisfy conservation of mass.

e. Solution and numerical techniques

The alternating direction implicit technique is used to achieve large time steps. Steps of 1–5 min have been achieved.

f. Boundary treatment

Lateral boundary conditions are obtained from a solution of 1D vertical equations. The top boundary is treated as a rigid lid.

g. Parameterization of subgrid mixing

Second-moment turbulent closures are used.

h. Cumulus parameterization

This is obtained by using a second-moment turbulent closure coupled with a Gaussian cloud model.

TABLE 12A.1. Examples of domains used in model runs.

x (km)	y (km)	z (km)
10	7	1
250	300	2
1600	1300	3
2000	1500	4

i. Surface boundary treatment

The surface boundary conditions are obtained by applying a heat energy balance at the surface, coupled with a soil layer, and tall tree canopy.

j. Special conditions

The model has been used to study diverse conditions including diurnal variation of PBL, lake breezes, transport and diffusion over complex terrain, industrial plumes, and the marine boundary layer.

k. Radiation

Solar and longwave radiation in the clouds are parameterized according to Hanson and Derr (1987). Longwave radiation for the ambient air is parameterized according to Sasamori (1968).

l. Stable precipitation

Precipitation microphysics are based on the work by Nickerson et al. (1986).

m. Algorithms to link to other models

Four-dimensional data assimilation and nested-grid capabilities have been developed.

n. Coding practices

The model is modular and has been installed on Sun, SGI, Data General, DEC, IBM, and Cray computers. It also runs on a laptop workstation.

Chapter 13

Statistical Evaluation of the Mesoscale Model Results

GÜNTER GROSS

Department of Meteorology, University of Hannover, Hannover, Germany

---

In chapters 9–12 results of simulations using phase I and phase II data from Project WIND are presented and particular aspects discussed for each of the models individually, that is, for FITNAH (chapter 9), Tel Aviv (chapter 10), RAMS (chapter 11), and HOTMAC (chapter 12). In this chapter the results are examined as a whole using the statistical approach suggested in chapter 6.

Calculated distributions in space and time, together with observations of the meteorological variables wind speed, wind direction, temperature, and downward solar radiation are available for comparison. The measures used here to compare the quality of the model results with the field observations are as follows: correlation coefficient R

$$R = \frac{\overline{(X_o - \bar{X}_o)(X_p - \bar{X}_p)}}{[\overline{(X_o - \bar{X}_o)^2}\,\overline{(X_p - \bar{X}_p)^2}]^{1/2}} \quad (13.1)$$

relative mean bias FB

$$FB = \frac{\bar{X}_o - \bar{X}_p}{0.5(\bar{X}_o + \bar{X}_p)} \quad (13.2)$$

percentage of cases within a factor of 2, FAC2

$$0.5 \leqslant \frac{X_o}{X_p} \leqslant 2 \quad (13.3)$$

mean difference MD

$$MD = \overline{X_o - X_p}, \quad (13.4)$$

where N is the number of observations, X_o represents the observed value of a variable, X_p the predicted value, and overbars represent averages in time (of hourly values) or in space over all data. The observed value is a 1-min average and the simulated one is a value at the time step closest to the hour.

In addition to the statistics (Tables 13.1–13.20), scatterplots and time series graphs are also displayed, the latter allowing intercomparison of the model performances. (Note that, in the plots of wind directions, 0° and 360° are the same.) In Tables 13.1–13.10 (phase I) station values are grouped into different ranges of terrain height (see Figs. 9.1 and 12.1 for key): namely,

level 1: valley bottom, around 100 m MSL (sites S1–S6);

level 2: east side of valley, around 500 m MSL (sites S7–S13 and BS2);

level 3: foothills on valley edge, 500–1000 m MSL (sites S14, C1, C4, C7, BC3);

level 4: mountains, around 1400 m MSL (sites C2, C3).

In Tables 13.11–13.20 (phase II) a smaller subset of these stations is used. In the scatterplots (Figs. 13.1–13.3 and 13.10–13.12) stations are separated into five elevation ranges, indicated by different markers.

137

TABLE 13.1. Correlation coefficient R for wind speed at 10 m AGL at the 21 surface stations.

Site	FITNAH	Tel Aviv	RAMS	HOTMAC
S1	0.26	0.67	0.27	0.85
S2	0.76	0.75	−0.15	0.88
S3	0.32	0.73	0.17	0.84
S4	0.16	0.65	−0.26	0.76
S5	0.13	−0.09	0.02	0.45
S6	0.41	0.36	0.62	0.60
S7	0.04	0.70	0.23	0.71
S8	0.46	0.70	0.06	0.73
S9	0.57	0.84	−0.15	0.90
S10	0.31	0.85	−0.16	0.87
S11	0.47	0.27	0.32	0.40
S12	0.09	0.81	0.25	0.89
S13	0.13	0.78	−0.17	0.89
BS2	−0.23	0.78	−0.01	0.70
S14	0.55	0.74	0.11	0.55
C1	0.40	0.93	0.49	0.83
C4	0.33	0.50	0.53	0.81
C7	0.39	0.83	0.42	0.83
BC3	0.73	0.62	−0.07	0.60
C2	0.71	0.20	0.24	0.80
C3	0.28	0.86	0.61	0.80
Mean	0.35	0.64	0.16	0.75

TABLE 13.2. As for Table 13.1 but for temperature at 2 m AGL.

Site	FITNAH	Tel Aviv	RAMS	HOTMAC
S1	0.97	0.97	0.72	0.99
S2	0.99	0.95	0.81	0.96
S3	0.97	0.97	0.72	0.98
S4	0.97	0.98	0.74	0.98
S5	0.93	0.84	0.88	0.83
S6	0.91	0.85	0.85	0.84
S7	0.84	0.66	0.87	0.68
S8	0.93	0.86	0.89	0.84
S9	0.95	0.87	0.82	0.90
S10	0.97	0.91	0.84	0.92
S11	0.93	0.81	0.83	0.83
S12	0.98	0.97	0.75	0.97
S13	0.98	0.96	0.77	0.96
BS2	0.95	0.88	0.82	0.91
S14	0.95	0.96	0.22	0.98
C1	0.97	0.97	0.52	0.98
C4	0.96	0.90	0.39	0.94
C7	0.93	0.83	0.52	0.86
BC3	0.96	0.97	0.54	0.98
C2	0.95	0.98	0.82	0.98
C3	0.96	0.98	0.87	0.99
Mean	0.95	0.91	0.72	0.92

TABLE 13.3. As for Table 13.1 but for downward solar radiation at ground level.

Site	FITNAH	Tel Aviv	RAMS	HOTMAC
S1	0.98	0.97	0.98	0.98
S2	0.99	0.99	0.99	0.99
S3	0.99	0.99	0.99	0.99
S4	0.98	0.98	0.98	0.99
S12	0.99	0.98	0.99	0.99
S13	0.99	0.97	0.99	0.99
S14	0.98	0.98	0.99	0.98
Mean	0.99	0.98	0.98	0.99

TABLE 13.4. Relative mean bias FB for wind speed
at 10 m AGL at the 21 surface stations.

Site	FITNAH	Tel Aviv	RAMS	HOTMAC
S1	0.16	0.53	−0.32	−0.14
S2	0.42	0.84	0.43	0.28
S3	0.34	0.66	−0.06	0.06
S4	0.36	0.64	0.15	0.00
S5	0.51	0.92	0.32	0.22
S6	0.29	0.71	0.06	−0.12
S7	0.64	0.79	0.83	0.44
S8	0.44	0.99	0.63	0.40
S9	0.32	0.85	0.56	0.28
S10	0.23	0.67	0.10	−0.02
S11	0.58	1.21	0.66	0.69
S12	0.27	0.69	0.00	−0.00
S13	0.51	0.79	0.19	0.17
BS2	0.60	0.85	0.25	0.26
S14	−0.45	0.43	−0.30	−0.23
C1	0.36	0.51	0.36	−0.13
C4	−0.19	0.40	0.16	0.08
C7	0.35	0.96	0.33	0.46
BC3	0.67	0.81	0.77	0.32
C2	0.04	0.43	−0.13	−0.68
C3	−0.06	−0.40	−0.39	−1.12

TABLE 13.5. As for Table 13.4 but for temperature at 2 m AGL.

Site	FITNAH	Tel Aviv	RAMS	HOTMAC
S1	−0.04	−0.07	−0.02	−0.21
S2	0.04	0.01	0.04	−0.12
S3	0.04	0.00	0.05	−0.14
S4	0.00	−0.04	0.01	−0.17
S5	0.06	0.02	0.06	−0.11
S6	0.08	0.04	0.09	−0.10
S7	0.05	0.03	0.07	−0.10
S8	0.08	0.04	0.10	−0.09
S9	0.07	0.05	0.10	−0.10
S10	0.03	0.01	0.06	−0.13
S11	0.08	0.06	0.08	−0.09
S12	−0.02	−0.05	−0.01	−0.19
S13	0.01	−0.03	0.00	−0.17
BS2	0.02	−0.01	0.01	−0.15
S14	0.01	−0.02	−0.00	−0.16
C1	0.04	−0.00	0.04	−0.12
C4	0.09	0.05	0.06	−0.07
C7	0.08	0.04	0.05	−0.10
BC3	0.01	0.26	0.07	−0.18
C2	−0.18	−0.15	−0.08	−0.33
C3	−0.21	−0.12	−0.15	−0.37

TABLE 13.6. Percentage of cases, where wind speed at 10 m AGL
is within a factor of 2 (FAC2) of the observed value.

Site	FITNAH	Tel Aviv	RAMS	HOTMAC
S1	68	56	40	76
S2	84	32	44	88
S3	56	32	44	72
S4	60	48	44	76
S5	60	24	60	64
S6	76	40	76	76
S7	36	40	28	68
S8	76	16	48	68
S9	72	36	44	68
S10	40	36	28	68
S11	44	20	60	52
S12	52	32	36	68
S13	68	36	20	60
BS2	52	28	44	56
S14	72	52	64	48
C1	72	64	60	84
C4	68	76	80	68
C7	64	8	72	64
BC3	60	24	36	80
C2	76	56	56	44
C3	40	40	60	8
Mean	61	37	49	64

TABLE 13.7. Mean difference MD for wind speed (m s^{-1}) at 10 m AGL at the 21 surface stations.

Site	FITNAH	Tel Aviv	RAMS	HOTMAC
S1	0.81	0.94	1.30	0.74
S2	1.09	1.76	1.74	0.87
S3	0.98	1.15	1.47	0.52
S4	1.06	1.21	1.71	0.76
S5	1.34	1.93	1.16	1.36
S6	0.78	1.07	0.78	1.07
S7	1.95	1.86	2.07	1.37
S8	1.24	1.94	1.46	1.28
S9	1.43	1.81	1.87	1.15
S10	0.95	1.15	1.36	0.71
S11	1.99	3.00	2.06	2.16
S12	1.29	1.35	1.46	0.73
S13	1.06	1.32	1.47	0.81
BS2	1.45	1.65	1.43	0.96
S14	0.99	0.80	0.96	0.88
C1	1.60	1.43	1.59	1.05
C4	0.84	0.94	0.78	0.93
C7	1.27	1.98	1.25	1.18
BC3	2.57	2.95	2.84	1.70
C2	1.17	1.54	1.50	1.82
C3	1.06	0.72	0.91	2.81
Mean	1.28	1.55	1.48	1.18

TABLE 13.8. As for Table 13.7 but for wind direction (°) at 10 m AGL.

Site	FITNAH	Tel Aviv	RAMS	HOTMAC
S1	42.41	55.73	36.44	70.05
S2	43.56	52.31	45.21	45.26
S3	49.49	61.17	37.14	48.80
S4	59.11	71.28	32.69	61.16
S5	50.17	44.22	38.44	39.28
S6	47.30	60.03	53.90	34.99
S7	71.33	49.36	71.61	46.31
S8	33.70	63.32	47.82	45.13
S9	41.48	51.76	48.00	45.89
S10	98.10	83.45	50.35	57.00
S11	41.43	62.75	46.84	58.18
S12	68.25	73.37	39.52	53.87
S13	62.20	54.63	33.80	49.06
BS2	65.47	72.87	62.38	69.15
S14	48.69	54.06	80.83	60.59
C1	65.32	44.01	60.44	60.91
C4	51.97	68.10	86.87	49.22
C7	45.82	50.21	91.88	50.05
BC3	35.36	53.62	56.83	37.21
C2	43.50	58.64	84.13	48.25
C3	88.11	92.19	76.46	63.76
Mean	54.89	60.81	56.27	52.10

TABLE 13.9. As for Table 13.7 but for temperature (K) at 2 m AGL.

Site	FITNAH	Tel Aviv	RAMS	HOTMAC
S1	2.10	2.16	3.64	6.11
S2	2.92	1.53	2.28	3.56
S3	1.75	2.58	4.58	4.69
S4	1.89	1.68	3.64	4.86
S5	3.56	2.24	1.98	3.30
S6	3.88	2.17	2.49	2.96
S7	4.58	3.37	2.11	2.94
S8	4.09	2.45	2.80	2.66
S9	3.50	1.94	2.58	3.06
S10	2.24	1.94	2.60	4.05
S11	3.82	2.72	2.34	2.65
S12	1.19	1.81	3.83	5.38
S13	1.51	1.83	3.60	4.99
BS2	2.56	2.08	2.50	4.19
S14	2.28	1.20	3.80	4.52
C1	1.70	1.64	4.66	3.51
C4	3.50	2.04	2.92	1.92
C7	3.68	2.42	2.28	2.92
BC3	2.89	5.35	2.86	4.75
C2	3.64	3.21	5.74	5.97
C3	4.03	4.54	6.50	6.90
Mean	2.92	2.42	3.32	4.09

TABLE 13.10. As for Table 13.7 but for downward solar radiation (W m^{-2}) at ground level.

Site	FITNAH	Tel Aviv	RAMS	HOTMAC
S1	81.02	123.12	80.60	72.47
S2	39.11	90.16	36.96	32.74
S3	39.51	88.40	36.60	33.80
S4	41.09	30.30	38.72	36.49
S12	27.68	85.96	30.16	18.63
S13	40.52	100.24	39.20	44.55
S14	35.13	90.44	34.56	37.23
Mean	43.44	86.95	42.40	39.41

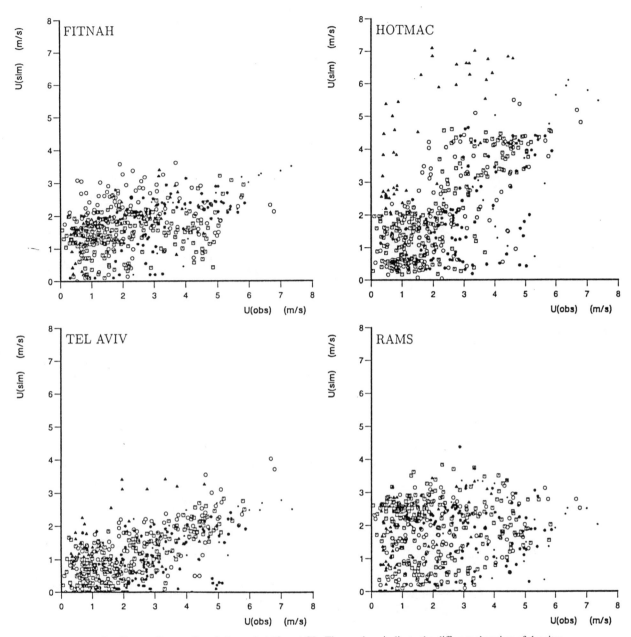

FIG. 13.1. Scatter diagram for wind speed at 10 m AGL. The markers indicate the different elevation of the sites (0–199 m, squares; 200–399 m, triangles; 400–599 m, asterisks; 600–999 m, circles; 1000–2000, dots).

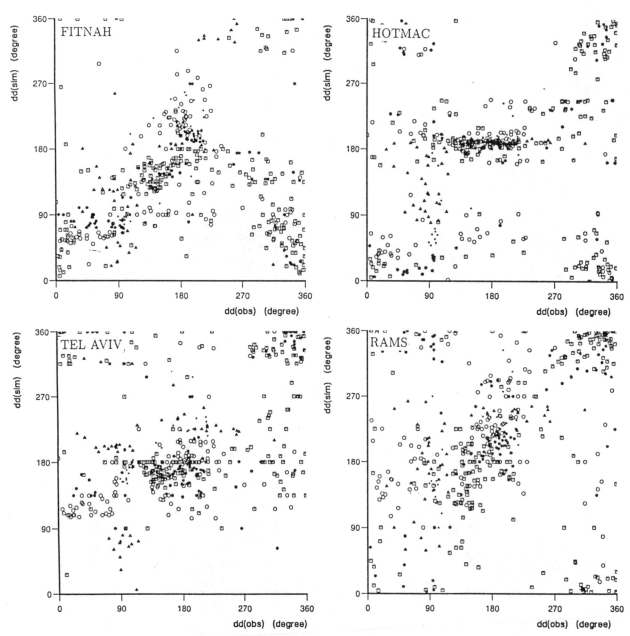

FIG. 13.2. As for Fig. 13.1 but for wind direction at 10 m AGL.

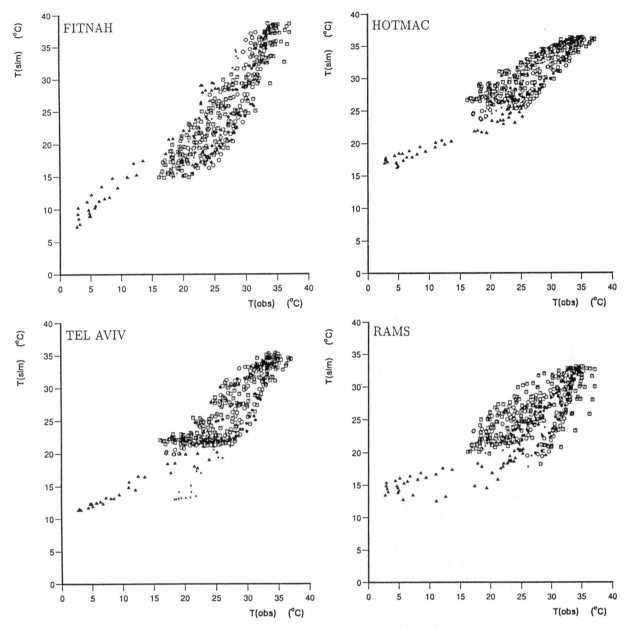

FIG. 13.3. As for Fig. 13.1 but for temperature at 2 m AGL.

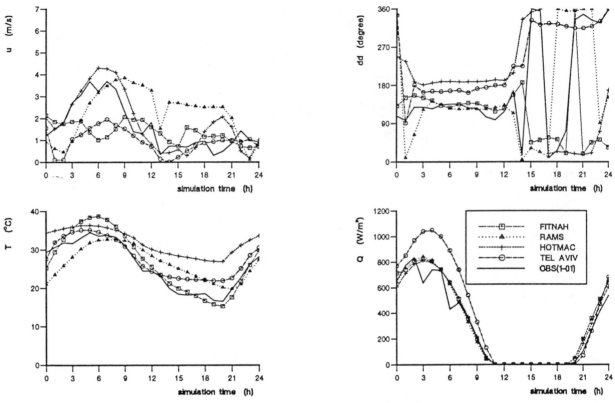

FIG. 13.4. Simulated and observed time variation of wind speed (10 m), wind direction (10 m), temperature (2 m), and downward solar radiation at ground level at station S1. The beginning of the simulation time is 0900 LST.

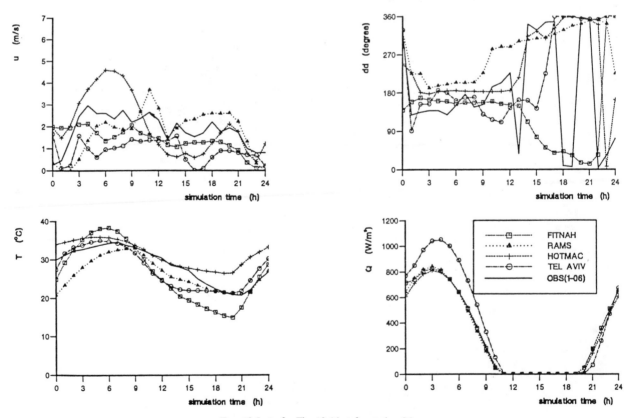

FIG. 13.5. As for Fig. 13.4 but for station S6.

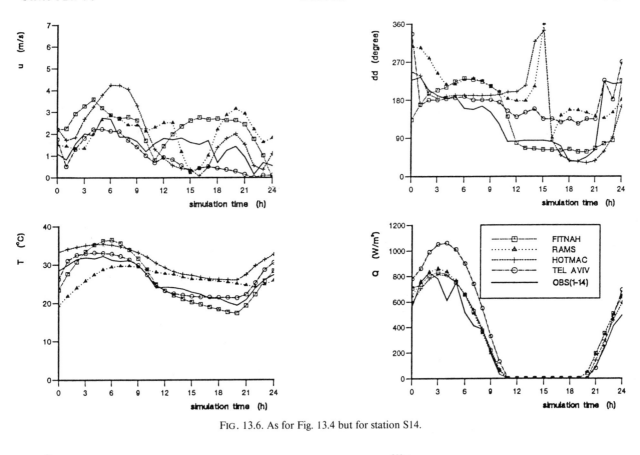

FIG. 13.6. As for Fig. 13.4 but for station S14.

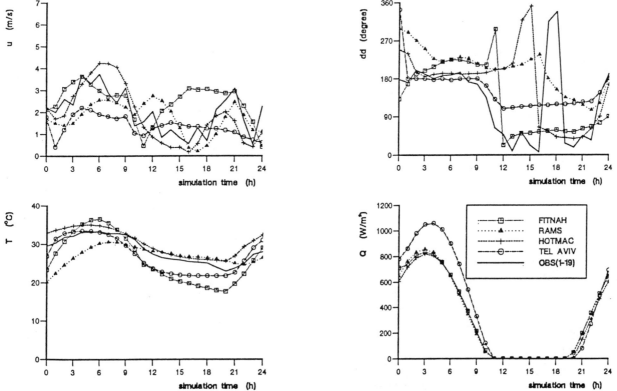

FIG. 13.7. As for Fig. 13.4 but for station C4.

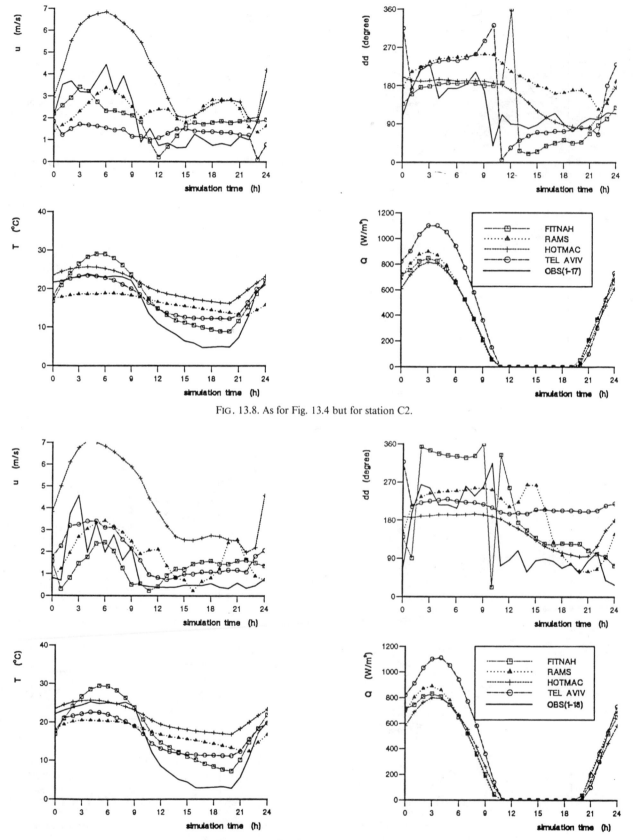

FIG. 13.8. As for Fig. 13.4 but for station C2.

FIG. 13.9. As for Fig. 13.4 but for station C3.

TABLE 13.11. Correlation coefficient R for wind speed at 10 m AGL at the 15 surface stations.

Site	FITNAH	Tel Aviv	RAMS	HOTMAC
S1	0.36	−0.38	0.72	0.33
S2	0.77	−0.31	0.57	0.21
S3	0.74	−0.37	0.64	0.26
S4	0.72	−0.46	0.80	0.39
S5	0.67	−0.41	0.66	0.47
S6	0.66	−0.53	0.66	0.60
S7	0.75	−0.29	0.86	0.83
S8	0.51	−0.74	0.87	0.75
S9	0.40	−0.68	0.91	0.76
S10	0.39	−0.64	0.87	0.65
S11	0.33	−0.50	0.91	0.80
S12	0.50	−0.51	0.75	0.45
S13	0.49	−0.09	0.61	0.37
BS2	0.22	0.17	0.74	0.45
S14	0.18	0.52	0.85	0.83
Mean	0.51	−0.35	0.76	0.54

TABLE 13.12. As for Table 13.11 but for temperature at 2 m AGL.

Site	FITNAH	Tel Aviv	RAMS	HOTMAC
S1	0.76	0.96	−0.58	0.95
S2	0.58	0.82	−0.29	0.78
S3	0.82	0.96	−0.60	0.94
S4	0.73	0.95	−0.40	0.93
S5	0.64	0.86	−0.36	0.79
S6	0.83	0.91	−0.55	0.87
S7	−0.26	−0.25	0.20	−0.26
S8	−0.15	−0.13	0.14	−0.15
S9	0.68	0.78	−0.44	0.76
S10	0.60	0.80	−0.37	0.76
S11	0.56	0.52	−0.42	0.58
S12	0.88	0.88	−0.65	0.93
S13	0.84	0.88	−0.62	0.92
BS2	0.79	0.79	−0.51	0.83
S14	0.77	0.85	−0.73	0.86
Mean	0.60	0.71	−0.41	0.70

TABLE 13.13. As for Table 13.11 but for downward solar radiation at ground level.

Site	FITNAH	Tel Aviv	RAMS	HOTMAC
S1	0.95	0.91	0.54	0.94
S2	0.87	0.84	0.66	0.88
S3	0.92	0.88	0.74	0.92
S4	0.90	0.76	0.83	0.90
S12	0.94	0.89	0.69	0.92
S13	0.92	0.92	0.68	0.91
S14	0.89	0.90	0.97	0.91
Mean	0.91	0.87	0.73	0.91

TABLE 13.14. Relative mean bias FB for wind speed at 10 m AGL at the 15 surface stations.

Site	FITNAH	Tel Aviv	RAMS	HOTMAC
S1	0.70	1.22	0.34	−0.20
S2	0.98	1.54	0.67	0.26
S3	0.81	1.30	0.41	−0.00
S4	0.89	1.33	0.45	0.11
S5	0.86	1.44	0.33	0.07
S6	0.62	1.28	0.02	−0.26
S7	1.13	1.47	0.63	0.31
S8	1.20	1.66	0.81	0.50
S9	1.14	1.60	0.70	0.38
S10	1.15	1.59	0.75	0.37
S11	1.16	1.42	0.86	0.38
S12	1.04	1.31	0.67	0.23
S13	1.02	1.21	0.71	0.22
BS2	1.06	1.13	0.80	0.23
S14	0.32	0.88	−0.02	−0.68

TABLE 13.15. As for Table 13.14 but for temperature at 2 m AGL.

Site	FITNAH	Tel Aviv	RAMS	HOTMAC
S1	−0.11	−0.13	−0.04	−0.01
S2	−0.19	−0.15	−0.08	−0.05
S3	−0.06	−0.11	0.01	0.05
S4	−0.04	−0.02	0.01	0.04
S5	−0.17	0.01	−0.07	−0.05
S6	−0.03	0.10	0.03	0.05
S7	−0.52	−0.52	−0.42	−0.37
S8	−0.14	−0.10	−0.33	−0.18
S9	−0.21	−0.25	−0.34	−0.35
S10	−0.52	−0.19	−0.43	−0.38
S11	−0.08	0.15	0.01	0.04
S12	0.08	0.21	0.17	0.20
S13	−0.08	0.04	0.06	−0.01
BS2	−0.06	0.05	0.07	0.01
S14	−0.14	0.00	0.01	−0.07

TABLE 13.16. Percentage of cases, where wind speed at 10 m AGL is within a factor of 2 (FAC2) of the observed value.

Site	FITNAH	Tel Aviv	RAMS	HOTMAC
S1	48	8	92	96
S2	16	4	36	88
S3	20	0	88	96
S4	12	0	96	100
S5	24	0	100	100
S6	48	12	100	100
S7	0	0	44	100
S8	0	0	16	88
S9	0	0	32	96
S10	0	0	24	96
S11	0	0	0	96
S12	0	0	36	100
S13	0	0	32	96
BS2	0	0	20	96
S14	88	16	100	40
Mean	17	2	54	92

TABLE 13.17. Mean difference MD for wind speed (m s⁻¹) at 10 m AGL at the 15 surface stations.

Site	FITNAH	Tel Aviv	RAMS	HOTMAC
S1	2.51	3.64	1.69	1.49
S2	5.23	6.82	4.17	2.59
S3	3.27	4.49	2.14	1.29
S4	3.99	5.16	2.46	1.60
S5	4.22	5.72	2.33	1.75
S6	2.25	3.69	1.08	1.75
S7	7.00	8.12	4.76	2.90
S8	8.26	9.92	6.40	4.54
S9	6.66	8.23	4.68	2.87
S10	7.02	8.45	5.30	3.13
S11	7.34	8.29	6.05	3.29
S12	5.36	6.19	3.96	2.01
S13	4.88	5.46	3.98	1.86
BS2	5.17	5.38	4.26	1.68
S14	1.09	2.10	0.73	3.61
Mean	4.95	6.11	3.60	2.42

TABLE 13.19. As for Table 13.17 but for temperature (K) at 2 m AGL.

Site	FITNAH	Tel Aviv	RAMS	HOTMAC
S1	1.07	0.50	1.43	0.65
S2	1.52	0.63	1.27	0.82
S3	0.86	0.60	1.45	0.70
S4	0.70	0.96	1.22	0.87
S5	0.84	0.84	1.33	1.11
S6	0.65	2.00	2.18	1.57
S7	2.96	2.55	1.79	2.84
S8	1.48	2.08	2.36	2.34
S9	0.55	1.40	1.83	1.33
S10	2.72	4.25	4.82	3.94
S11	0.86	1.39	1.17	1.43
S12	0.74	0.66	1.42	0.84
S13	0.79	0.75	1.54	0.78
BS2	0.67	1.84	1.98	1.52
S14	2.18	0.94	1.16	1.78
Mean	1.24	1.43	1.80	1.50

TABLE 13.18. As for Table 13.17 but for wind direction (°) at 10 m AGL.

Site	FITNAH	Tel Aviv	RAMS	HOTMAC
S1	18.93	31.97	13.77	14.73
S2	11.35	53.30	8.71	11.43
S3	15.15	43.88	8.26	9.52
S4	16.22	48.23	9.41	10.03
S5	14.79	20.93	12.03	21.14
S6	20.89	30.82	13.79	12.72
S7	14.40	21.87	7.32	15.21
S8	20.83	23.78	5.60	9.96
S9	30.95	38.79	13.27	8.74
S10	22.49	31.83	8.66	7.69
S11	32.14	31.41	11.91	9.49
S12	15.85	30.99	6.63	11.82
S13	14.72	19.13	7.33	11.02
BS2	14.45	16.90	9.09	11.26
S14	29.13	16.72	13.22	13.61
Mean	19.48	30.70	9.93	11.89

TABLE 13.20. As for Table 13.17 but for downward solar radiation (W m⁻²) at ground level.

Site	FITNAH	Tel Aviv	RAMS	HOTMAC
S1	46.59	53.64	78.12	48.20
S2	59.36	67.20	71.48	57.46
S3	55.86	46.44	39.92	59.52
S4	61.90	72.12	60.52	63.79
S12	39.78	56.92	67.88	49.44
S13	48.41	59.44	66.00	50.68
S14	61.33	24.68	20.04	69.65
Mean	53.32	54.35	57.71	56.96

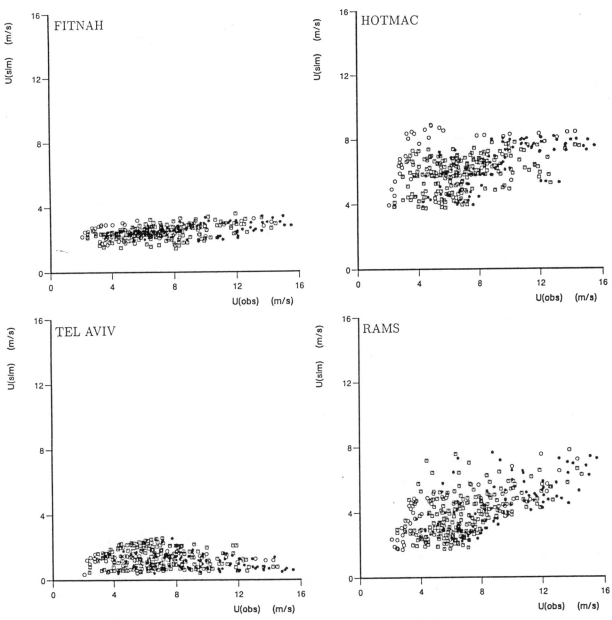

FIG. 13.10. Scatter diagram for wind speed at 10 m AGL. The markers indicate the different elevation of the sites (0–199 m, squares; 200–399 m, triangles; 400–599 m, asterisks; 600–999 m, circles; 1000–2000, dots).

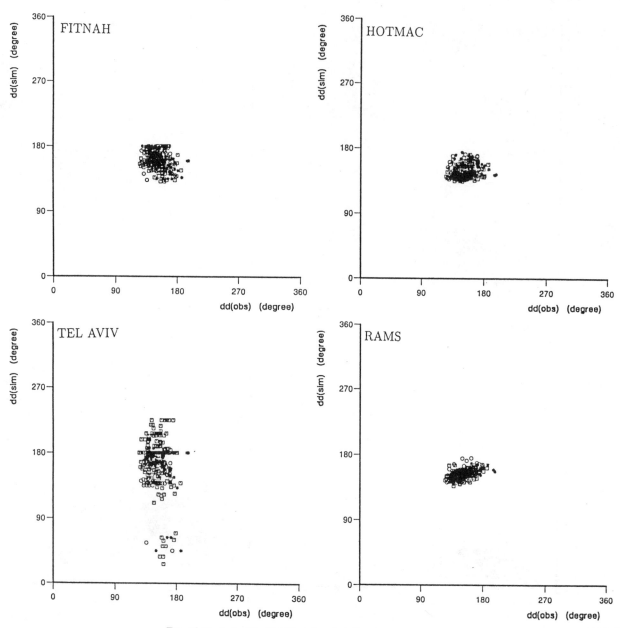

FIG. 13.11. As for Fig. 13.10 but for wind direction at 10 m AGL.

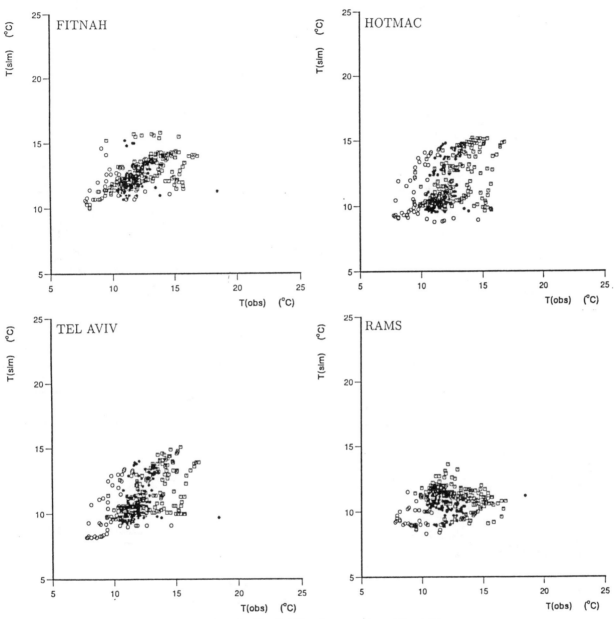

FIG. 13.12. As for Fig. 13.10 but for temperature at 2 m AGL.

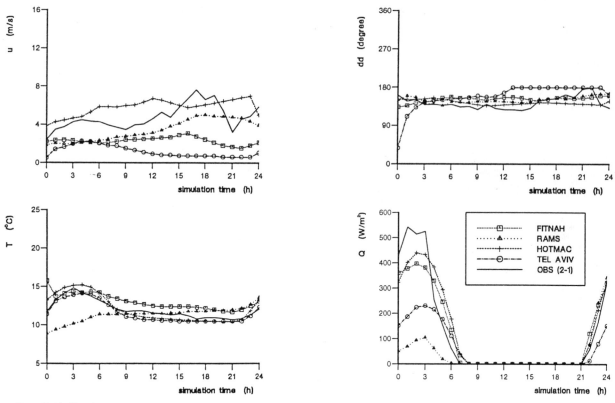

FIG. 13.13. Simulated and observed time variation of wind speed (10 m), wind direction (10 m), temperature (2 m), and downward solar radiation at ground level at station S1. The beginning of the simulation time is 1000 LST.

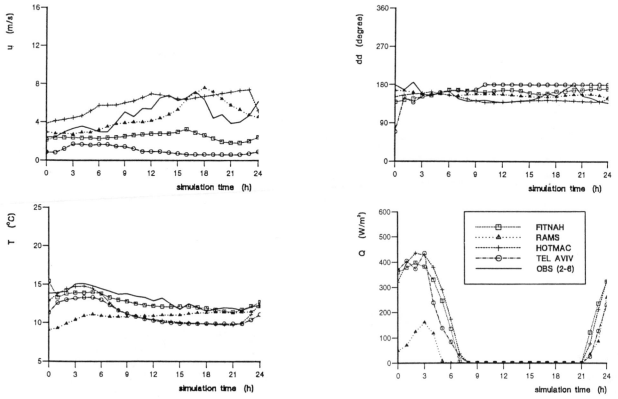

FIG. 13.14. As for Fig. 13.13 but for station S6.

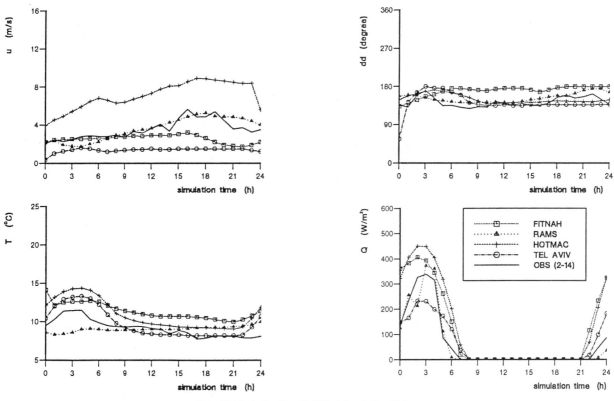

FIG. 13.15. As for Fig. 13.13 but for station S14.

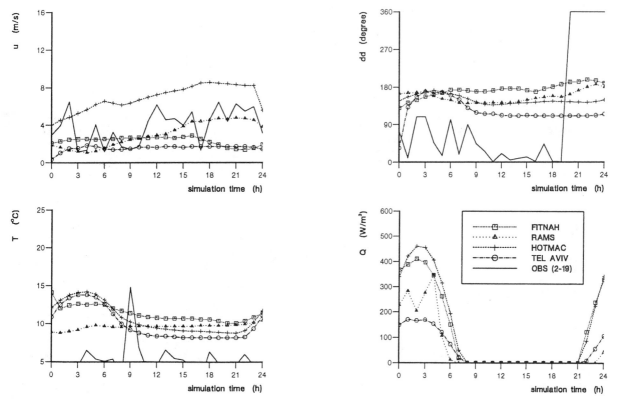

FIG. 13.16. As for Fig. 13.13 but for station C4.

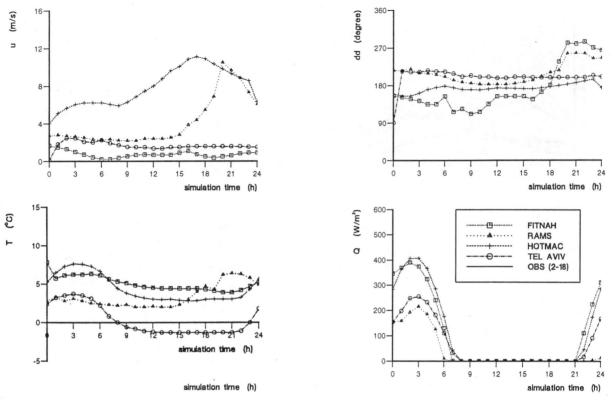

FIG. 13.17. As for Fig. 13.13 but for station C2.

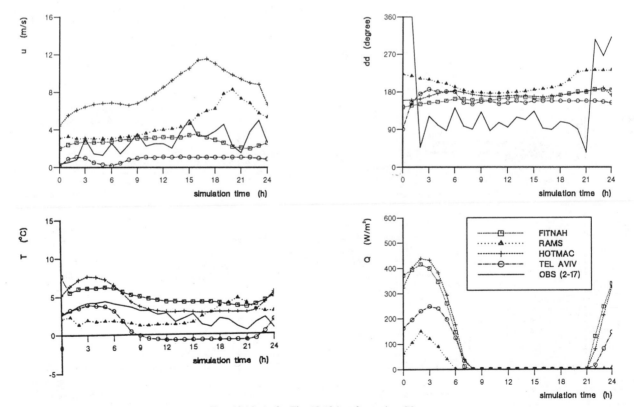

FIG. 13.18. As for Fig. 13.13 but for station C3.

Chapter 14

Comments on Statistical Results

NIELS E. BUSCH

Danish Technological Institute, Copenhagen, Denmark

WERNER KLUG

Department of Meteorology, Technical University of Darmstadt, Darmstadt, Germany

ROBERT P. PEARCE

Department of Meteorology, University of Reading, Reading, United Kingdom

PETER WHITE

Meteorological Office, Bracknell, Berkshire, United Kingdom

14.1. Introduction

The following must be borne in mind.

• The model outputs are volume averages on a horizontal 5 km × 5 km grid, whereas the observations with which they are compared are point values that may in reality differ considerably from the averages. The observation sites were far from homogeneous on the grid scale, with differing terrain heights, slope inclinations, land usage, and soil conditions all of particular importance under the conditions of strong thermal forcing in phase I.

• The rapidly changing synoptic conditions of phase II, including a frontal passage, were not adequately imposed, through either the lateral boundary conditions or internally, in any of the models. The RAMS model, using grid nesting, made the most ambitious attempt to represent these conditions. Even in this case, however, the verification region, with its high terrain, seemed not to be able to adjust sufficiently rapidly to the imposed larger-scale changes.

• The modelers were free to decide their own initialization and data assimilation procedures, so that, inevitably, the results represent differing levels of sophistication of external forcing.

• Some modelers were able to carry out a series of experiments to determine the most appropriate values of some assigned model parameters—for example, surface albedo and ground wetness—before generating the results published here. The results of such experiments are, in these cases, discussed in the text. They throw useful light on the physics of the modeled phenomena, particularly for phase I; only in the case of

the RAMS model was such an approach adopted for phase II. Those results reported here that have not been preceded by such experiments must therefore not be regarded as reflecting the full potential of the models used.

14.2. Commentary

Considerable encouragement can be derived from the ability of all the models to reproduce, with good accuracy, the diurnally forced upslope and downslope winds of phase I, including their times of onset. Also, even for phase II, the nudging procedure adopted by HOTMAC seems, after experiments to determine the optimum values of the nudging parameters, to have been reasonably successful in retaining the broad-scale features of the observed changes.

Some interesting similarities and differences in the performances of the models emerge including the following.

• In phase I, the low nighttime temperatures observed at 2 m over the high ground were not predicted by any of the models (Figs. 13.8 and 13.9). No doubt two of the factors contributing to this were 1) failure of the models to resolve a strong surface inversion and 2) the differences between the model (5 km × 5 km average) terrain heights and the actual station heights.

• The models failed to reproduce the range of observed phase II (and, to a lesser extent, phase I) wind speeds, the FITNAH and Tel Aviv models completely so (Fig. 13.10). This was no doubt largely due to their inability to resolve the rapid changes associated with the frontal passage as discussed above; also, in the case of the Tel Aviv model, only a couple of experiments were run—these with a most basic initialization. On the other hand, all of the models, apart from the Tel Aviv model, reproduced the narrow range of phase II wind directions, the RAMS model being particularly successful (Fig. 13.11).

• All the models, apart from RAMS in phase II, were reasonably successful in predicting temperature changes, apart from those over the high ground during the night in phase I (Figs. 13.3 and 13.12, Tables 13.2, 13.5, 13.9, 13.12, 13.15, and 13.19). The reason for this deficiency of the RAMS model is not clear.

• The scatter diagram of wind directions for phase I (Fig. 13.2) shows a model bias toward a direction of 180° in the plot for HOTMAC and, to a lesser extent, in the plot for the Tel Aviv model.

• HOTMAC was the least successful of the models in predicting the phase I nighttime cooling, not only at the hill stations, but also in the valley (Figs. 13.4 and 13.5). The reason for this was that the thermal diffusivity and specific heat capacity of soil used in the simulations were too large. (Additional simulations carried out after the project using smaller values for these parameters resulted in temperature predictions that were in better agreement with the observations than those shown here.)

It is clear from these few selected examples that each model has been able to demonstrate some of its own particular strengths and weaknesses; none is clearly better or worse than the others overall. Each of the modeling groups involved is able to use these results to prepare its own strategy for model development. It is with this in mind that the U.S. Army has been encouraged to organize a future workshop based on experimentation with data from phases III and IV of Project WIND in addition to further experiments with the data from phases I and II.

REFERENCES

REFERENCES

Achtemeier, G. L., 1975: On the initialization problem: A variational adjustment method. *Mon. Wea. Rev.,* **103,** 1089–1103.

——, 1989: Modification of a successive corrections objective analysis for improved derivative calculations. *Mon. Wea. Rev.,* **117,** 78–86.

Air Resources Laboratory, 1989: Across North America Tracer Experiment (ANATEX). Vol. 1: Description, primary site ground-level sampling, and meteorology. NOAA Tech. Memo., ERL ARL-167, 83 pp. [Available from NOAA, Air Resources Laboratory, Silver Spring, MD 20910.]

Alpert, P., and B. Getenio, 1988a: One-level diagnostic modeling of mesoscale surface winds in complex terrain. Part I: Comparison with three-dimensional modeling in Israel. *Mon. Wea. Rev.,* **116,** 2025–2046.

——, and ——, 1988b: One-level diagnostic modeling of mesoscale surface winds in complex terrain. Part II: Applicability of short-range forecasting. *Mon. Wea. Rev.,* **116,** 2047–2061.

André, J. C., G. De Moor, P. Lacarrère, G. Therry, and R. du Vachat, 1978: Modeling the 24-hour evolution of the mean and turbulent structures of the planetary boundary layer. *J. Atmos. Sci.,* **35,** 1861–1883.

——, P. Bougeault, J.-F. Mahfouf, P. Mascart, J. Noilhan, and J.-P. Pinty, 1989: Impact of forests on mesoscale meteorology. *Philos. Trans. Roy. Soc. London,* **324,** 407–422.

Andrén, A., 1990: Evaluation of a turbulence closure scheme suitable for air-pollution applications. *J. Appl. Meteor.,* **29,** 224–239.

——, 1991: A TKE-dissipation model for the atmospheric boundary layer. *Bound.-Layer Meteor.,* **56,** 207–221.

Anthes, R. A., 1974: Data assimilation and initialization of hurricane prediction models. *J. Atmos. Sci.,* **31,** 702–719.

——, 1977: A cumulus parameterization scheme utilizing a one-dimensional cloud model. *Mon. Wea. Rev.,* **105,** 270–286.

——, 1984: Predictability of mesoscale meteorological phenomena. *Proc. Institute of Physics Conf.,* La Jolla, CA, American Institute of Physics, 247–270.

——, 1990: Recent applications of the Penn State/NCAR mesoscale model to synoptic, mesoscale and climate studies. *Bull. Amer. Meteor. Soc.,* **71,** 1610–1629.

——, and T. T. Warner, 1978: Development of hydrodynamic models suitable for air pollution and other mesometeorological studies. *Mon. Wea. Rev.,* **106,** 1045–1078.

——, and D. P. Baumhefner, 1984: A diagram depicting forecast skill and predictability. *Bull. Amer. Meteor. Soc.,* **65,** 701–703.

——, E.-Y. Hsie, and Y.-H. Kuo, 1987: Description of the Penn State/NCAR Mesoscale Model version 4 (MM4). NCAR Tech. Note, TN-282+STR, 66 pp. [Available from National Center for Atmospheric Research, P.O. Box 3000, Boulder, CO 80307.]

——, Y.-H. Kuo, E.-Y. Hsie, S. Low-Nam, and T. W. Bettge, 1989: Estimation of skill and uncertainty in regional numerical models. *Quart. J. Roy. Meteor. Soc.,* **115,** 763–806.

Arakawa, A., and W. H. Schubert, 1974: Interaction of a cumulus cloud ensemble with the large-scale environment. Part I. *J. Atmos. Sci.,* **31,** 674–701.

——, and J.-M. Chen, 1987: Closure assumptions in the cumulus parameterization problem. *Short and Medium Range Numerical Weather Prediction,* T. Matsuno, Ed., Universal Academy Press, 107–131.

Arritt, R. W., 1987: The effect of water surface temperature on lake breeze and thermal internal boundary layers. *Bound.-Layer Meteor.,* **40,** 101–125.

Artz, R., R. A. Pielke, and J. Galloway, 1985: Comparison of the ARL/ATAD constant level and the NCAR isentropic trajectory analysis for selected case studies. *Atmos. Environ.,* **19,** 47–63.

Atwater, M. A., and P. S. Brown, 1974: Numerical computations of the latitudinal variation of solar radiation for an atmosphere of varying opacity. *J. Appl. Meteor.,* **13,** 289–297.

Avissar, R., and R. A. Pielke, 1989: A parameterization of heterogeneous land surfaces for atmospheric numeric models and its impact on regional meteorology. *Mon. Wea. Rev.,* **117,** 2113–2136.

Baer, F., 1977: Adjustment of initial conditions required to suppress gravity oscillations in nonlinear flows. *Contrib. Atmos. Phys.,* **50,** 350–366.

Ballard, S. P., B. W. Golding, and R. N. B. Smith, 1991: Mesoscale model experimental forecasts of the Haar of northeast Scotland. *Mon. Wea. Rev.,* **119,** 2107–2123.

Barchet, W. R., 1987: Evaluation of regional-scale air quality models. Chairmen reports from four workshops, EPRI-EA-5473, Electric Power Research Institute, Palo Alto, CA, 130 pp.

——, and R. L. Dennis, 1989: The NAPAP model evaluation, Vol. 1: Protocol. U.S. Environmental Protection Agency, AREAL, MD-80, Research Triangle Park, NC, 94 pp.

Barker, E., S. G. Benjamin, F. H. Carr, J. Derber, G. DiMego, J. A. McGinley, R. O. McPherson, R. A. Petersen, T. W. Schlatter, and N. Seaman, 1994: Status of critical activities in mesoscale data assimilation for the U.S. Weather Research Program. *Bull. Amer. Meteor. Soc.,* submitted.

Barnes, S. L., 1973: Mesoscale objective map analysis using weighted time-series observations. NOAA Tech. Memo., ERL NSSL-62, National Severe Storms Laboratory, Norman, OK, 60 pp. [NTIS COM-73-10781.]

Beck, M. B., 1987: Water quality modeling: A review of the analysis of uncertainty. *Water Resour. Res.,* **23,** 1393–1442.

Beljaars, A. C. M., and A. A. M. Holstag, 1991: Flux parameterization over land surfaces for atmospheric modes. *J. Appl. Meteor.,* **30,** 327–341.

——, J. L. Walmsley, and P. A. Taylor, 1987: A mixed spectral finite-difference model for neutrally stratified boundary-layer flow over roughness changes and topography. *Bound.-Layer Meteor.,* **38,** 273–303.

Benjamin, S. G., 1983: Some effects of surface heating and topography on the regional severe storm environment. Ph.D. thesis, The Pennsylvania State University, 265 pp.

——, 1989: An isentropic mesoα-scale analysis system and its sensitivity to aircraft and surface observations. *Mon. Wea. Rev.,* **117,** 1586–1603.

——, and N. L. Seaman, 1985: A simple scheme for objective analysis in curved flow. *Mon. Wea. Rev.,* **113,** 1184–1198.

Benniston, M. G., and G. Sommeria, 1981: Use of a detailed planetary boundary layer model for parameterization purposes. *J. Atmos. Sci.,* **32,** 780–797.

Benoit, R., and M. Roch, 1987: Impact of a direct assimilation of satellite-inferred precipitation rates into a regional scale meteorological model. *Proc. Symp. Mesoscale Analysis and Forecasting,* Vancouver, British Columbia, Canada, IAMAP, ESA SP-282, 617–623.

——, J. Cote, and J. Mailhot, 1989: Inclusion of a TKE boundary layer parameterization in the Canadian regional finite-element model. *Mon. Wea. Rev.,* **117,** 1726–1750.

Bergthorsson, P., and B. R. Doos, 1955: Numerical weather map analysis. *Tellus,* **7,** 329–340.

Berkowicz, R., and L. P. Prahm, 1979: Generalization of K-theory for turbulent diffusion. Part I: Spectral turbulent diffusivity concept. *J. Appl. Meteor.,* **18,** 255–272.

——, and ——, 1982: Evaluation of the profile method for estimation of surface fluxes of momentum and heat. *Atmos. Environ.,* **16,** 2809–2819.

Betts, A. K., 1975: Parametric interpretation of trade-wind cumulus budget studies. *J. Atmos. Sci.,* **32,** 1934–1945.

——, 1986: A new convective adjustment scheme. I: Observational and theoretical basis. *Quart. J. Roy. Meteor. Soc.,* **112,** 677–692.

Blackadar, A. K., 1962: The vertical distribution of wind and turbulent exchange in a neutral atmosphere. *J. Geophys. Res.,* **67,** 3095–3102.

——, 1979: High-resolution models of the planetary boundary layer. Vol. 1, *Advances in Environmental Science and Engineering,* 50–85.

Bleck, R., 1975: An economical approach to the use of wind data in the optimum interpolation of geo- and Montgomery potential fields. *Mon. Wea. Rev.,* **103,** 807–816.

Bodin, S., 1980: Applied numerical modeling of the atmospheric boundary layer. *Atmospheric Planetary Boundary Layer Physics,* A. Longhetto, Ed., Elsevier, 1–76.

Bornstein, R. D., 1975: The two-dimensional URBMET urban boundary layer model. *J. Appl. Meteor.,* **14,** 1459–1477.

Bougeault, P., 1985: A simple parameterization of the large scale effects of convection. *Mon. Wea. Rev.,* **113,** 2108–2121.

——, and J.-C. André, 1986: On the stability of the third-order turbulence closure for the modeling of the stratocumulus-topped boundary layer. *J. Atmos. Sci.,* **43,** 1574–1581.

——, and P. Lacarrère, 1989: Parameterization of orography-induced turbulence in a mesobeta-scale model. *Mon. Wea. Rev.,* **117,** 1872–1890.

Bourke, W., and J. L. McGregor, 1983: A nonlinear vertical mode initialization scheme for a limited area prediction model. *Mon. Wea. Rev.,* **111,** 2285–2297.

Bouttier, F., J.-F. Mahfouf, and J. Noilhan, 1993a: Sequential assimilation of soil moisture from low-level atmospheric parameters. Part I: Sensitivity and calibration studies. *J. Appl. Meteor.,* **32,** 1335–1351.

——, ——, and ——, 1993b: Sequential assimilation of soil moisture from low-level atmospheric parameters. Part II: Implementation in a mesoscale model. *J. Appl. Meteor.,* **32,** 1352–1364.

Bratseth, A. M., 1982: A simple and efficient approach to the initialization of weather prediction models. *Tellus,* **34,** 352–357.

——, 1986: Statistical interpolation by means of successive corrections. *Tellus,* **38A,** 439–447.

Brier, G. W., 1990: A historical and personal perspective of model evaluations in meteorology. *Bull. Amer. Meteor. Soc.,* **71,** 349–351.

Briere, S., 1982: Nonlinear normal mode initialization scheme of a limited area model. *Mon. Wea. Rev.,* **110,** 1166–1186.

——, 1987: Energetics of daytime sea breeze circulation as determined from a two-dimensional third-order turbulence closure model. *J. Atmos. Sci.,* **44,** 1455–1474.

Brost, R. A., and J. C. Wyngaard, 1978: A model study of the stably stratified planetary boundary layer. *J. Atmos. Sci.,* **35,** 1427–1440.

——, P. L. Haagenson, and Y.-H. Kuo, 1988: Eulerian simulation of tracer distribution during CAPTEX. *J. Appl. Meteor.,* **27,** 579–593.

Bush, N., S. W. Chang, and R. A. Anthes, 1976: A multi-level model of the planetary boundary layer suitable for use with mesoscale dynamic models. *J. Appl. Meteor.,* **15,** 909–919.

Businger, J. A., J. C. Wyngaard, Y. Izumi, and E. F. Bradley, 1971: Flux-profile relationships in the atmospheric surface layer. *J. Atmos. Sci.,* **28,** 181–189.

Buyn, D. W., 1990: On the analytical solution of flux-profile relationships in the atmospheric surface layer. *J. Appl. Meteor.,* **29,** 652–657.

Carlson, T. N., and F. H. Ludlam, 1968: Conditions for the occurrence of severe local storms. *Tellus,* **20,** 203–226.

——, S. G. Benjamin, G. S. Forbes, and Y.-F. Li, 1983: Elevated mixed layers in the regional severe storms environment: Con-

ceptual model and case studies. *Mon. Wea. Rev.,* **111,** 1453–1473.

Carr, F. H., and L. F. Bosart, 1978: A diagnostic evaluation of rainfall predictability for Tropical Storm Agnes, June 1972. *Mon. Wea. Rev.,* **106,** 363–374.

——, and M. E. Baldwin, 1991: Incorporation of observed precipitation estimates during the initialization of synoptic and mesoscale storms. *First Inter. Winter Storm Symp.,* New Orleans, LA, Amer. Meteor. Soc., 71–75.

Chang, J. S., R. A. Brost, L. S. A. Isaksen, S. Madronich, P. Middleton, W. R. Stockwell, and C. J. Walcek, 1987: A three-dimensional Eulerian acid deposition model. Physical concepts and formulation. *J. Geophys. Res.,* **92,** 14 681–14 700.

Chock, D. P., and Y.-H. Kuo, 1990: Comparison of wind-field models using the CAPTEX data. *J. Appl. Meteor.,* **29,** 76–91.

Cionco, R. M., 1984: Test plan for Project WIND. U.S. Army Atmospheric Sciences Laboratory, White Sands Missile Range, NM.

——, 1985: Modeling airflow over variable terrain. *Proc. HAZMAT '85 West Conf.,* Long Beach, CA.

——, 1987: Windfield simulations from a hierarchy of nested meso- and micrometeorological models. *Proc. Symp. Mesoscale Analysis and Forecasting,* Vancouver, British Columbia, Canada, IUGG.

——, 1989: AMADEUS: A dispersion study over moderately complex terrain. *Proc. Sixth Joint Conf. on Applications of Air Pollution Meteorology (with APCA).* Anaheim, CA, Amer. Meteor. Soc., 260–263.

——, 1991: Project WIND documentation and user guide, Phase II: 24-hr period, 1000 HRS PST, 1 February–1000 HRS PST, 2 February 1986. U.S. Army Atmospheric Sciences Laboratory, White Sands Missile Range, NM.

Clark, T. L., 1979: Numerical simulations with a three-dimensional cloud model: Lateral boundary experiments and multi-cellular severe storm simulations. *J. Atmos. Sci.,* **36,** 2191–2215.

Claussen, M., 1991: Estimation of areally-averaged surface fluxes. *Bound.-Layer Meteor.,* **54,** 387–410.

Cohen, C., and W. M. Frank, 1987: Simulation of tropical convective systems. Part II: Simulations of moving cloud lines. *J. Atmos. Sci.,* **44,** 3800–3820.

Collins, W. G., and L. S. Gandin, 1990: Comprehensive hydrostatic quality control at the National Meteorological Center. *Mon. Wea. Rev.,* **118,** 2752–2767.

Costigan, K. R., and W. R. Cotton, 1992: Large eddy simulations of the atmospheric boundary layer east of the Colorado Rockies. *Tenth Symp. on Turbulence and Diffusion,* Portland, OR, Amer. Meteor. Soc., J461–J463.

Cotton, W. R., 1986: Averaging and the parameterization of physical processes in mesoscale models. *Mesoscale Meteorology and Forecasting,* P. S. Ray, Ed., Amer. Meteor. Soc., 614–635.

——, and R. A. Anthes, 1989: *Storm and Cloud Dynamics,* Academic Press, 883 pp.

——, M. A. Stephens, T. Nekhorn, and G. J. Tripoli, 1982: The Colorado State University three-dimensional cloud/mesoscale model. Part II: An ice phase parameterization. *J. Rech. Atmos.,* **16,** 296–320.

——, C. J. Tremback, and R. L. Walko, 1988: CSU RAMS—A cloud model goes regional. *Proc. NCAR Workshop on Limited-Area Modeling Intercomparison,* Boulder, CO, NCAR, 202–211.

Courtier, P., and O. Talagrand, 1987: Variational assimilation of meteorological observations with the adjoint vorticity equation. Part 2: Numerical results. *Quart. J. Roy. Meteor. Soc.,* **113,** 1329–1368.

Cox, W. W., and J. A. Tikvart, 1990: Statistical procedure for determining the best performing air quality simulation model. *Atmos. Environ.,* **24,** 2387–2395.

Cram, J. M., R. A. Pielke, and W. R. Cotton, 1992a: Numerical simulation and analysis of a prefrontal squall line. Part I: Observations and basic simulation results. *J. Atmos. Sci.,* **49,** 189–208.

——, ——, and ——, 1992b: Numerical simulation and analysis of a prefrontal squall line. Part II: Propagation of the squall line as an internal gravity wave. *J. Atmos. Sci.,* **49,** 209–225.

Cressman, G., 1959: An operational objective analysis system. *Mon. Wea. Rev.,* **87,** 367–374.

Daley, R., 1981: Normal mode initialization. *Rev. Geophys. Space Phys.,* **19,** 450–468.

——, 1991: *Atmospheric Data Analysis.* Cambridge University Press, 457 pp.

Danard, M., 1985: On the use of satellite estimates of precipitation in initial analyses for numerical weather prediction. *Atmos.-Ocean.,* **23,** 23–42.

Davies, H. C., 1976: A lateral boundary formulation for multi-level prediction models. *Quart. J. Roy. Meteor. Soc.,* **102,** 405–418.

——, 1983: Limitations of some common lateral boundary schemes used in regional NWP models. *Mon. Wea. Rev.,* **111,** 1002–1012.

——, and R. E. Turner, 1977: Updating prediction models by dynamical relaxation: An examination of the technique. *Quart. J. Roy. Meteor. Soc.,* **103,** 225–245.

Deardorff, J. W., 1966: The counter-gradient heat flux in the lower atmosphere and in the laboratory. *J. Atmos. Sci.,* **23,** 503–506.

——, 1972: Parameterization of the planetary boundary layer for use in general circulation models. *Mon. Wea. Rev.,* **100,** 93–106.

Dennis, R. L., and S. K. Seilkop, 1986: The use of spatial patterns and uncertainty estimates in the model evaluation process. *Proc. Fifth Joint Conf. on Applications of Air Pollution Meteorology,* Boston, MA, Amer. Meteor. Soc., 41–44.

——, W. R. Barchet, T. L. Clark, S. K. Seilkop, and P. M. Roth, 1989: Evaluation of regional acid deposition models. State-of-Science/Tech. Rep. No. 5, National Acid Precipitation Assessment Program, USEPA, AREAL, Mb-80, Research Triangle Park, NC, 157 pp.

Deterling, H. W., and D. Etling, 1985: Application of the E-ϵ turbulence closure model to the atmospheric boundary layer. *Bound.-Layer Meteor.,* **33,** 113–133.

DiMego, G. J., 1988: The National Meteorological Center regional analysis system. *Mon. Wea. Rev.,* **116,** 977–1000.

Donaldson, C. D., 1973: Construction of a dynamic model of the production of atmospheric turbulence and the dispersal of atmospheric pollutants. *Workshop on Micrometeorology,* Boston, MA, Amer. Meteor. Soc., 313–392.

Donner, L. J., 1988: An initialization for cumulus convection in numerical weather prediction models. *Mon. Wea. Rev.,* **116,** 377–385.

——, and P. J. Rasch, 1989: Cumulus initialization in a global model for numerical weather prediction models. *Mon. Wea. Rev.,* **117,** 2653–2671.

Douglas, S. G., and R. C. Kessler, 1991: Analysis of mesoscale airflow patterns in the south-central coast air basin during the SCCCAMP 1985 intensive measurement periods. *J. Appl. Meteor.,* **30,** 607–631.

Draxler, R. R., 1979: Modeling the results of two recent mesoscale dispersion experiments. *Atmos. Environ.,* **13,** 1523–1533.

——, 1982: Measuring and modeling the transport and dispersion of krypton-85 1500 km from a point source. *Atmos. Environ.,* **16,** 2763–2776.

——, 1987: Sensitivity of a trajectory model to the spatial and temporal resolution of the meteorological data during CAPTEX. *J. Climate Appl. Meteor.,* **26,** 1577–1588.

——, 1990: The calculation of low-level winds from the archived data of a regional primitive equation model. *J. Appl. Meteor.,* **29,** 240–248.

——, R. Dietz, R. J. Lagomarsino, and G. Start, 1991: Across North America Tracer Experiment (ANATEX): Sampling and analysis. *Atmos. Environ.,* **25A,** 2815–2836.

Dudek, M., 1988: Numerical study of the formation of mesoscale convective complexes. Ph.D. dissertation, State University of New York at Albany, 266 pp.

Dudhia, J., 1989: Numerical study of convection observed during the winter monsoon experiment using a mesoscale, two-dimensional model. *J. Atmos. Sci.,* **46,** 3077–3107.

Durran, D. R., and J. B. Klemp, 1983: A compressible model for the simulation of moist mountain waves. *Mon. Wea. Rev.,* **111,** 2341–2361.

Dyer, A. J., and B. B. Hicks, 1970: Flux-gradient relationships in the constant flux layer. *Quart. J. Roy. Meteor. Soc.,* **96,** 715–721.

Ebert, E. E., U. Schumann, and R. B. Stull, 1989: Nonlocal turbulent mixing in the convective boundary layer evaluated from large-eddy simulation. *J. Atmos. Sci.,* **46,** 2178–2207.

Eddy, A., 1964: The objective analysis of horizontal wind divergence fields. *Quart. J. Roy. Meteor. Soc.,* **90,** 424–440.

Eliassen, A., 1954: Provisional report on calculation of spatial covariance and autocorrelation of the pressure field. *Dynamic Meteorology: Data Assimilation Methods,* L. Bengtsson, M. Ghil, and E. Kallen, Eds., Springer-Verlag, 319–330.

Ellingson, R. G., and Y. Fouquart, 1991: The intercomparison of radiation codes in climate models: An overview. *J. Geophys. Res.,* **96**(D5), 8925–8928.

——, J. Ellis, and S. Fels, 1991: The intercomparison of radiation codes in climate models: Longwave results. *J. Geophys. Res.,* **96**(D5), 8929–8954.

Emanuel, K. A., 1983: On assessing the local conditional symmetric instability from atmospheric soundings. *Mon. Wea. Rev.,* **111,** 2016–2033.

——, 1991: A scheme for representing cumulus convection in large-scale models. *J. Atmos. Sci.,* **48,** 2313–2335.

Enger, L., 1986: A higher order closure model applied to dispersion in a convective PBL. *Atmos. Environ.,* **20,** 879–894.

——, 1990: Simulation of dispersion in moderately complex terrain. Part A: The fluid dynamic model. *Atmos. Environ.,* **24A,** 2431–2446.

Errico, R. M., 1990: An analysis of dynamic balance in a mesoscale model. *Mon. Wea. Rev.,* **118,** 558–572.

——, and D. Baumhefner, 1987: Predictability experiments using a high-resolution limited area model. *Mon. Wea. Rev.,* **115,** 488–504.

——, and P. J. Rasch, 1988: A comparison of various normal mode initialization schemes and the inclusion of diabatic processes. *Tellus,* **40A,** 1–25.

——, and D. L. Williamson, 1988: The behavior of gravitational modes in numerical forecasts with the NCAR Community Climate Model. *Mon. Wea. Rev.,* **116,** 1737–1756.

Estoque, M. A., 1968: Vertical mixing due to penetrative convection. *J. Atmos. Sci.,* **25,** 561–575.

Ferber, G. J., J. L. Heffter, R. R. Draxler, R. J. Lagomarsino, F. L. Thomas, R. N. Dietz, and C. M. Benkovitz, 1986: Cross-Appalachian Tracer Experiment (CAPTEX '83) final report. NOAA Tech. Memo., ERL AFL-142, 60 pp. [Available from NOAA, Air Resources Laboratory, Silver Spring, MD 20910.]

Fiedler, B. H., 1984: An integral closure model for the vertical turbulent flux of a scalar in a mixed layer. *J. Atmos. Sci.,* **41,** 674–680.

Flatau, P. J., 1985: Study of second-order turbulence closure technique and its application to atmospheric flows. Atmospheric Science paper 393. [Available from Dept. of Atmospheric Science, Colorado State University, Fort Collins, CO 80523.]

Fouquart, Y., J. C. Buriez, M. Herman, and R. S. Kandel, 1990: The influence of clouds on radiation: A climate-modeling perspective. *Rev. Geophys.,* **28,** 145–166.

——, B. Bonnel, and V. Ramaswamy, 1991: Intercomparing shortwave radiation codes used in climate models. *J. Geophys. Res.,* **96**(D5), 8955–8968.

Fox, D. G., 1981: Judging air quality model performance—Review of the Woods Hole workshop. *Bull. Amer. Meteor. Soc.,* **62,** 599–609.

Frank, W. M., 1983: The cumulus parameterization problem. *Mon. Wea. Rev.,* **111,** 1859–1871.

——, 1984: A cumulus parameterization scheme incorporating subgrid scale convective forcing. Postprints, *15th Conf. on Hurri-*

canes and Tropical Meteorology, Miami, FL, Amer. Meteor. Soc., 183–190.

——, and C. Cohen, 1987: Simulation of tropical convective systems. Part I: A cumulus parameterization. *J. Atmos. Sci.,* **44,** 3787–3799.

Fritsch, J. M., and C. F. Chapell, 1980a: Numerical prediction of convectively driven mesoscale pressure systems. Part I: Convective parameterization. *J. Atmos. Sci.,* **37,** 1722–1733.

——, and ——, 1980b: Numerical prediction of convectively driven mesoscale pressure systems. Part II: Mesoscale model. *J. Atmos. Sci.,* **37,** 1734–1762.

Gabriel, K. R., 1987: Some reflections on the role of statisticians in weather experiments. Preprints, *Tenth Conf. on Probability and Statistics in Atmospheric Sciences,* Boston, MA, Amer. Meteor. Soc., J31–J34.

Gadd, A., 1978: A split explicit integration scheme for numerical weather prediction. *Quart. J. Roy. Meteor. Soc.,* **104,** 569–582.

Galperin, B., L. H. Kantha, S. Hassid, and A. Rosati, 1988: A quasi-equilibrium turbulent energy model for geophysical flows. *J. Atmos. Sci.,* **45,** 55–62.

Gandin, L. S., 1965: *Objective Analysis of Meteorological Fields.* Israel Program for Scientific Translations, 242 pp. [NTIS N66-18047.]

——, 1988: Complex quality control of meteorological observations. *Mon. Wea. Rev.,* **116,** 1137–1156.

Gerrity, J., 1977: The LFM model—1976: A documentation. NOAA Tech. Memo. NWS NMC-60.

Ghil, M., S. Cohn, J. Tavantzis, K. Bube, and E. Issacson, 1981: Applications of estimation theory to numerical weather prediction. *Dynamic Meteorology: Data Assimilation Methods,* L. Bengtsson, M. Ghil, and E. Kallen, Eds., Springer-Verlag, 139–224.

Gierens, K. M., 1993: A fast six-flux radiative transfer model for application in finite cloud models. *Beitr. Phys. Atmos.,* **66,** 73–88.

Golding, B. W., 1987: Strategies for using mesoscale data in an operational mesoscale model. *Proc. Symp. Mesoscale Analysis and Forecasting,* Vancouver, British Columbia, Canada, ESA, SP-282, 569–578.

——, and S. Ballard, 1989: Stratiform cloud precipitation. Met O 11 Mesoscale Documentation Paper No. 8, Meteorological Office, United Kingdom.

Goody, R. M., and Y. L. Yung, 1989: *Atmospheric Radiation.* Oxford University Press. 519 pp.

Grell, G. A., 1993: Prognostic evaluation of assumptions used by cumulus parameterizations. *Mon. Wea. Rev.,* **121,** 764–787.

——, Y.-H. Kuo, and R. J. Pasch, 1991: Semiprognostic tests of cumulus parameterization schemes in the midlatitudes. *Mon. Wea. Rev.,* **119,** 5–31.

Gronas, S., and K. H. Midtbo, 1986: Operational multivariate analyses by successive corrections. *J. Meteor. Soc. Japan,* **64,** 61–74.

Gross, G., 1990: On the wind field in the Loisach Valley—Numerical simulation and comparison with the LOWEX III data. *Meteor. Atmos. Phys.,* **42,** 231–247.

Haagenson, P. L., Y.-H. Kuo, M. Skumanich, and N. L. Seaman, 1987: Tracer verification of trajectory models. *J. Climate Appl. Meteor.,* **26,** 410–426.

——, K. Gao, and Y.-H. Kuo, 1990: Evaluation of meteorological analyses, simulations, and long-range transport calculations using ANATEX surface tracer data. *J. Appl. Meteor.,* **29,** 1268–1283.

Hadfield, M. G., W. R. Cotton, and R. A. Pielke, 1992: Large-eddy simulations of thermally forced circulations in the convective boundary layer. Part II: The effect of changes in wavelength and wind speed. *Bound.-Layer Meteor.,* **58,** 307–327.

Haltiner, G. J., and R. T. Williams, 1980: *Numerical Prediction and Dynamic Meteorology.* John Wiley and Sons, 477 pp.

Hanna, S. R., 1988: Air quality model evaluation and uncertainty. *J. Air Pollut. Control Assoc.,* **38,** 406–412.

——, 1989: Confidence limits for air quality models, as estimated by bootstrap and jackknife resampling methods. *Atmos. Environ.,* **23,** 1385–1395.

——, and J. C. Chang, 1992: Representativeness of wind measurements on a mesoscale grid with station separations of 312 m to 10 km. *Bound.-Layer Meteor.,* **60,** 309–324.

——, D. G. Strimaitis, J. S. Scire, G. E. Moore, and R. C. Kessler, 1991: Overview of results of analysis of data from the South-Central Coast Cooperative Aerometric Monitoring Program (SCCCAMP 1985). *J. Appl. Meteor.,* **30,** 511–533.

Hansen, J. E., and L. D. Travis, 1974: Light scattering in planetary atmospheres. *Space Sci. Rev.,* **16,** 527–610.

Hanson, H. P., and V. E. Derr, 1987: Parameterization of radiative flux profiles within layer clouds. *J. Climate Appl. Meteor.,* **11,** 1511–1521.

Harms, D. E., S. Raman, and R. V. Madala, 1992a: An examination of four-dimensional data-assimilation techniques for numerical weather prediction. *Bull. Amer. Meteor. Soc.,* **73,** 425–440.

——, K. D. Sashegyi, R. V. Madala, and S. Raman, 1992b: Four-dimensional data assimilation of GALE data using a multivariate analysis scheme and a mesoscale model with diabatic initialization. NRL Tech. Memo., No. 7147, Naval Research Laboratory, Washington, DC, 236 pp. [NTIS A256063.]

Hasse, L., 1993: Turbulence closure in boundary-layer theory—An invitation to debate. *Bound.-Layer Meteor.,* **65,** 249–254.

Hassid, S., and B. Galperin, 1983: A turbulent energy model for geophysical flow. *Bound.-Layer Meteor.,* **26,** 397–412.

Hayden, C. M., 1973: Experiments in the four-dimensional assimilation of *NIMBUS 4* SIRS data. *J. Appl. Meteor.,* **12,** 425–436.

Helfand, H. M., and J. C. Labraga, 1988: Design of a nonsingular level 2.5 second-order closure model for the prediction of atmospheric turbulence. *J. Atmos. Sci.,* **45,** 113–132.

Herzegh, P. H., and A. R. Jameson, 1992: Observing precipitation through dual-polarization radar measurements. *Bull. Amer. Meteor. Soc.,* **73,** 1365–1374.

Hobbs, P. V., and A. Deepak, Eds., 1981: *Clouds, Their Function, Optical Properties, and Effects.* Academic Press, 497 pp.

Hoecker, W. H., 1977: Accuracy of various techniques for estimating boundary-layer trajectories. *J. Appl. Meteor.,* **16,** 374–383.

Hoke, J. E., and R. A. Anthes, 1976: The initialization of numerical modes by a dynamic-initialization technique. *Mon. Wea. Rev.,* **104,** 1551–1556.

Hosker, R. P., Jr., C. J. Nappo, Jr., and S. P. Hanna, 1974: Diurnal variation of vertical thermal structure in a pine plantation. *Agric. Meteor.,* **13,** 259–265.

Houze, R. A., 1989: Observed structure of mesoscale convective systems and implications for large-scale heating. *Quart. J. Roy. Meteor. Soc.,* **115,** 425–462.

Hsie, E.-Y., 1987: MM4 (Penn State/NCAR) Mesoscale Model Version 4 Documentation. NCAR Tech. Note, NCAR/TN-294+STR, 215 pp. [Available from National Center for Atmospheric Research, P.O. Box 3000, Boulder, CO 80307.]

——, and R. A. Anthes, 1984: Simulations of frontogenesis in a moist atmosphere using alternative parameterizations of condensation and precipitation. *J. Atmos. Sci.,* **41,** 2701–2716.

Huang, C.-Y., and S. Raman, 1989: Application of the *e–ε* closure model to simulations of mesoscale topographic effects. *Bound.-Layer Meteor.,* **49,** 169–195.

Johnson, D. E., P. K. Wang, and J. M. Straka, 1993: The role of ice-phase physics in a highly glaciated northern High Plains supercell storm. *J. Appl. Meteor.,* **32,** 745–759.

Johnson, R. H., 1976: The role of convective-scale precipitation downdrafts in cumulus and synoptic-scale interactions. *J. Atmos. Sci.,* **33,** 1890–1910.

Jones, R. W., 1986: Mature structure and motion of a model tropical cyclone with latent heating by the resolvable scales. *Mon. Wea. Rev.,* **114,** 973–990.

Juvanon du Vachat, R., 1986: A general formulation of normal modes for limited-area models: Application to initialization. *Mon. Wea. Rev.,* **114,** 2478–2487.

Kahl, J. D., and P. J. Samson, 1986: Uncertainty in trajectory calculations due to low-resolution meteorological data. *J. Climate Appl. Meteor.,* **25,** 1816–1831.

——, and ——, 1988: Trajectory sensitivity to rawinsonde data resolution. *Atmos. Environ.,* **22,** 1291–1299.

Kalnay, E., and A. Dalcher, 1987: Forecasting forecast skill. *Mon. Wea. Rev.,* **115,** 349–356.

Kao, C. Y. J., and T. Yamada, 1988: Use of the CAPTEX data for evaluations of a long-range transport numerical technique. *Mon. Wea. Rev.,* **116,** 293–306.

Keeler, R. J., B. W. Lewis, and G. R. Gray, 1989: Description of NCAR/FOF CP-2 meteorological Doppler radar. Preprints, *24th Conf. on Radar Meteorology,* Tallahassee, FL, Amer. Meteor. Soc., 589–592.

Kessler, R. C., L. L. Schuman, S. G. Douglas, and E. L. Hoving, 1989: Analysis of wind fields for the SCCCAMP 1985 intensive measurement periods. MM3-89-0034, Minerals Management Service, U.S. Department of the Interior, Pacific OCS Region, Los Angeles, CA, 416 pp.

Keyser, D., and R. A. Anthes, 1977: The applicability of a mixed-layer model of the planetary boundary layer to real-data forecasting. *Mon. Wea. Rev.,* **105,** 1351–1370.

Kimura, F., 1989: Heat flux on mixtures of different land-use surface: Test of a new parameterization scheme. *J. Meteor. Soc. Japan,* **67,** 401–408.

Kistler, R. E., and R. D. McPherson, 1975: On the use of a local wind correction technique in four-dimensional data assimilation. *Mon. Wea. Rev.,* **103,** 445–449.

Kitada, T., 1987: Effect of non-zero divergence wind fields on atmospheric transport calculations. *Atmos. Environ.,* **21,** 785–788.

Kitade, T., 1983: Non-linear normal mode initialization with physics. *Mon. Wea. Rev.,* **111,** 2194–2213.

Klemp, J. B., and D. K. Lilly, 1978: Numerical simulation of hydrostatic mountain waves. *J. Atmos. Sci.,* **35,** 78–107.

——, and R. B. Wilhelmson, 1978: The simulation of three-dimensional convective storm dynamics. *J. Atmos. Sci.,* **35,** 1070–1096.

Klug, W., G. Graziani, G. Grippa, D. Pierce, and C. Tassone, 1992: Evaluation of long term atmospheric transport models using environmental radioactivity data from the Chernobyl accident. *The ATMES Report,* Elsevier, 366 pp.

Koch, S. E., M. DesJardins, and P. J. Kocin, 1983: An interactive Barnes objective map analysis scheme for use with satellite and conventional data. *J. Climate Appl. Meteor.,* **22,** 1487–1503.

Kondratyev, K. Ya., 1969: *Radiation in the Atmosphere.* Academic Press, 912 pp.

Kreitzberg, C., and D. Perkey, 1976: Release of potential instability. Part I: A sequential plume model within a hydrostatic primitive equation model. *J. Atmos. Sci.,* **33,** 456–475.

——, and ——, 1977: Release of potential instability. Part I: The mechanism of convective mesoscale interaction. *J. Atmos. Sci.,* **34,** 1569–1595.

Krishnamurti, T. N., and W. J. Moxim, 1971: On parameterization of convective and nonconvective latent heat release. *J. Appl. Meteor.,* **10,** 3–13.

——, K. Ingles, S. Cooke, T. Kitade, and R. Pasch, 1984: Details of low-latitude, medium-range numerical weather prediction using a global spectral model. Part II: Effects of orography and physical initialization. *J. Meteor. Soc. Japan,* **62,** 163–648.

——, H. S. Bedi, W. Heckley, and K. Ingles, 1988: Reduction of the spinup time for evaporation and precipitation in a spectral model. *Mon. Wea. Rev.,* **109,** 1914–1929.

——, J. Xue, H. S. Bedi, K. Ingles, and D. Oosterhof, 1990: Physical initialization for numerical weather prediction over the tropics. FSU Report No. 90-5, Department of Meteorology, The Florida State University.

Krueger, S. K., 1988: Numerical simulation of tropical cumulus clouds and their interaction with the subcloud layer. *J. Atmos. Sci.,* **45,** 2221–2250.

Kuo, H. L., 1965: On formation and intensification of tropical cyclones through latent heat release by cumulus convection. *J. Atmos. Sci.,* **22,** 40–63.

——, 1974: Further studies of the parameterization of the influence of cumulus convection of the large scale flow. *J. Atmos. Sci.,* **31,** 1232–1240.

——, and S. Low-Nam, 1990: Prediction of nine explosive cyclones over the western North Atlantic Ocean with a regional model. *Mon. Wea. Rev.,* **118,** 3–25.

Kuo, Y.-H., and Y.-R. Guo, 1989: Dynamic initialization using observations from a network of profilers. *Mon. Wea. Rev.,* **117,** 1975–1998.

——, M. Skumanich, P. L. Haagenson, and J. S. Chang, 1985: The accuracy of trajectory models as revealed by the observing system simulation experiments. *Mon. Wea. Rev.,* **113,** 1852–1867.

Kurihara, Y., 1973: A scheme for moist convective adjustment. *Mon. Wea. Rev.,* **101,** 547–553.

Kyle, T. G., 1991: *Atmospheric Transmission, Emission, and Scattering.* Pergamon Press, 287 pp.

Lakhtakia, M. L., and T. T. Warner, 1987: A real-data numerical study of the development of precipitation along the edge of an elevated mixed layer. *Mon. Wea. Rev.,* **115,** 156–168.

Lamb, R. G., 1981: A numerical investigation of tetroon versus fluid particle dispersion in the convective planetary boundary layer. *J. Appl. Meteor.,* **20,** 391–403.

Lee, T. J., R. A. Pielke, T. G. F. Kittel, and J. F. Weaver, 1993: Atmospheric modeling and its spatial representation of land surface characteristics. *Integrating Geographic Information Systems and Environmental Modeling,* M. Goodchild, B. Parks, and L. T. Steyaert, Eds., Oxford University Press 108–122.

Leith, C., 1980: Nonlinear normal mode initialization and quasigeostrophic theory. *J. Atmos. Sci.,* **37,** 958–968.

Lejanas, H., 1979: Initialization of moisture in primitive equation models. *Mon. Wea. Rev.,* **88,** 1299–1305.

Lenoble, J., Ed., 1985: *Radiative Transfer in Scattering and Absorbing Atmospheres: Standard Computational Procedures.* A. Deepak, 300 pp.

Lewellen, W. S., and R. I. Sykes, 1989: Meteorological data needs for modeling air quality uncertainties. *J. Atmos. Oceanic Technol.,* **6,** 759–768.

Lilly, D. K., 1960: On the theory of disturbances in a conditionally unstable atmosphere. *Mon. Wea. Rev.,* **88,** 1–17.

——, 1990: Numerical prediction of thunderstorms—Has its time come? *Quart. J. Roy. Meteor. Soc.,* **116,** 779–798.

Lin, Y.-L., R. D. Farley, and H. D. Orville, 1983: Bulk parameterization of the snow field in a cloud model. *J. Climate Appl. Meteor.,* **22,** 1065–1092.

Linke, F., 1970: *Meteorologisches Taschenbuch,* II. Band. Akedemische Verlagsgesellschaft.

Liou, K.-N., 1980: *An Introduction to Atmospheric Radiation.* International Geophysics Series, Vol. 26, Academic Press, 392 pp.

Lipton, A. E., and T. H. Vonder Haar, 1990a: Mesoscale analysis by numerical modeling coupled with sounding retrieval from satellites. *Mon. Wea. Rev.,* **118,** 1308–1329.

——, and ——, 1990b: Preconvective mesoscale analysis over irregular terrain with a satellite-model coupled system. *Mon. Wea. Rev.,* **118,** 1330–1358.

Łobocki, L., 1992: Mellor–Yamada simplified second-order closure models: Analysis and application of the generalized von Kármán local similarity hypothesis. *Bound.-Layer Meteor.,* **59,** 83–109.

——, 1993: A procedure for the derivation of surface-layer bulk relationships from the simplified second-order closure models. *J. Appl. Meteor.,* **32,** 126–138.

Lord, S. J., 1982: Interaction of a cumulus cloud ensemble with the large-scale environment. Part III: Semi-prognostic tests of the Arakawa–Schubert cumulus parameterization. *J. Atmos. Sci.,* **39,** 88–103.

Lorenc, A. C., 1981: A global three-dimensional multivariate statistical interpolation scheme. *Mon. Wea. Rev.,* **109,** 701–721.

——, 1986: Analysis methods for numerical weather prediction. *Quart. J. Roy. Meteor. Soc.,* **112,** 1177–1194.

——, and O. Hammon, 1988: Objective quality control of observations using Bayesian methods. Theory, and a practical implementation. *Quart. J. Roy. Meteor. Soc.,* **114,** 515–543.

——, R. S. Bell, and B. MacPherson, 1991: The Meteorological Office analysis correction data assimilation scheme. *Quart. J. Roy. Meteor. Soc.,* 117, 59–89.

Louis, J. F., 1979: A parametric model of vertical eddy fluxes in the atmosphere. *Bound.-Layer Meteor.,* 17, 187–202.

Ly, L. N., 1991: An application of the e–ϵ turbulence model for studying coupled air–sea boundary layer structure. *Bound.-Layer Meteor.,* 54, 327–346.

Lynch, P., 1985a: Initialization using Laplace transforms. *Quart. J. Roy. Meteor. Soc.,* 111, 243–258.

——, 1985b: Initialization of a barotropic limited-area model using the Laplace transform technique. *Mon. Wea. Rev.,* 113, 1338–1344.

Machenhauer, B., 1977: On the dynamics of gravity oscillations in a shallow water model, with applications to normal mode initialization. *Contrib. Atmos. Phys.,* 50, 253–271.

Mahrer, Y., and R. A. Pielke, 1977a: A numerical study of the airflow over irregular terrain. *Contrib. Atmos. Phys.,* 50, 98–113.

——, and ——, 1977b: The effects of topography on sea and land breezes in a two-dimensional numerical model. *Mon. Wea. Rev.,* 105, 1151–1162.

——, and ——, 1978: A test of an upstream spline interpolation technique for the advective terms in a numerical mesoscale model. *Mon. Wea. Rev.,* 106, 818–830.

Maddox, R. A., 1980: An objective technique for separating macroscale and mesoscale features in meteorological data. *Mon. Wea. Rev.,* 108, 1108–1121.

Manabe, S., J. Smagorinsky, and R. F. Strickler, 1965: Simulated climatology of a general circulation model with a hydrological cycle. *Mon. Wea. Rev.,* 93, 769–798.

Maryon, R. H., and C. C. Heasman, 1988: The accuracy of plume trajectories forecast using the U.K. Meteorological Office operational forecasting models and their sensitivity to calculation schemes. *Atmos. Environ.,* 22, 259–272.

McDonald, J. E., 1960: Direct absorption of solar radiation by atmospheric water vapour. *J. Meteor.,* 17, 319–328.

McNider, R. T., and R. A. Pielke, 1981: Diurnal boundary layer development over sloping terrain. *J. Atmos. Sci.,* 38, 2198–2212.

Meador, W. E., and W. R. Weaver, 1980: Two-stream approximations to radiative transfer in planetary atmospheres: A unified description of existing methods and a new improvement. *J. Atmos. Sci.,* 37, 630–643.

Mellor, G. L., 1973: Analytic predictions of the properties of stratified planetary surface layers. *J. Atmos. Sci.,* 30, 1061–1069.

——, and T. Yamada, 1974: A hierarchy of turbulence closure models for planetary boundary layers. *J. Atmos. Sci.,* 31, 1791–1806. [Corrigendum. *J. Atmos. Sci.,* 34, 1482.]

——, and ——, 1982: Development of a turbulence closure model for geophysical fluid problems. *Rev. Geophys. Space Phys.,* 20, 851–875.

Miller, M. J., and A. J. Thorpe, 1981: Radiation conditions for the lateral boundaries of limited-area models. *Quart. J. Roy. Meteor. Soc.,* 107, 615–628.

Mills, G. A., and R. S. Seaman, 1990: The BMRC regional data assimilation system. *Mon. Wea. Rev.,* 118, 1217–1237.

Mintz, Y., and G. K. Walker, 1993: Global fields of soil and land surface evapotranspiration derived from observed precipitation and surface air temperature. *J. Appl. Meteor.,* 32, 1305–1334.

Miyakoda, K., and R. Moyer, 1968: A method of initialization for dynamical weather forecasting. *Tellus,* 20, 115–128.

——, J. Smagorinsky, R. F. Strickler, and G. D. Hembree, 1969: Experimental extended predictions with a nine-level hemispheric model. *Mon. Wea. Rev.,* 97, 1–76.

Molinari, J., 1982: Numerical hurricane prediction using assimilation of remotely-sensed rainfall rates. *Mon. Wea. Rev.,* 110, 553–571.

——, and T. Corsetti, 1985: Incorporation of cloud-scale and mesoscale downdrafts into a cumulus parameterization: Results of one- and three-dimensional integrations. *Mon. Wea. Rev.,* 113, 485–501.

——, and M. Dudek, 1992: Parameterization of convective heating in numerical weather prediction models. *Mon. Wea. Rev.,* 114, 1822–1831.

Mosteller, F., and J. W. Tukey, 1977: *Data Analysis and Regression.* Addison-Wesley, 588 pp.

Murphy, A. H., and R. L. Winkler, 1987: A general framework for forecast verification. *Mon. Wea. Rev.,* 115, 1330–1338.

——, and E. S. Epstein, 1989: Skill scores and correlation coefficients in model verification. *Mon. Wea. Rev.,* 117, 572–581.

Nickerson, E. C., E. Richard, R. Rosset, and D. R. Smith, 1986: The numerical simulation of clouds, rain, and air flow over the Vosges and Black Forest Mountains: A meso-β-model with parameterized microphysics. *Mon. Wea. Rev.,* 114, 398–414.

Nieuwstadt, F. T. M., 1984: The turbulent structure of the stable, nocturnal boundary layer. *J. Atmos. Sci.,* 41, 2202–2216.

Nitta, T., and J. Hovermale, 1969: A technique of objective analysis and initialization for the primitive forecast equations. *Mon. Wea. Rev.,* 97, 652–658.

Nordeng, T. E., 1987: The effect of vertical and slantwise convection in the simulation of polar lows. *Tellus,* 39A, 354–375.

O'Brien, J. J., 1970: A note on vertical structure of the eddy exchange coefficient in the planetary boundary layer. *J. Atmos. Sci.,* 27, 1213–1215.

Oliver, H. R., 1971: Wind profiles in and above a forest canopy. *Quart. J. Roy. Meteor. Soc.,* 97, 548–553.

Ookouchi, Y., M. Segal, R. C. Kessler, and R. A. Pielke, 1984: Evaluation of soil moisture effects on the generation and modification of mesoscale circulations. *Mon. Wea. Rev.,* 112, 2281–2292.

Ooyama, K. V., 1971: A theory on parameterization of cumulus convection. *J. Meteor. Soc. Japan,* 49, 744–756.

——, 1982: Conceptual evolution of the theory and modeling of the tropical cyclone. *J. Meteor. Soc. Japan,* 60, 369–379.

Orlanski, I., 1975: A rational subdivision of scales for atmospheric processes. *Bull. Amer. Meteor. Soc.,* 56, 527–530.

——, 1976: A simple boundary condition for unbounded hyperbolic flows. *J. Comput. Phys.,* 21, 251–269.

Pack, D..H., G. J. Ferber, J. L. Heffter, K. Telegadas, J. K. Angell, W. H. Hoecker, and L. Machta, 1978: Meteorology of long-range transport. *Atmos. Environ.,* 12, 425–444.

Paltridge G. W., and C. M. R. Platt, 1981: Aircraft measurements of solar and infrared radiation and the microphysics of cirrus cloud. *Quart. J. Meteor. Soc.,* 107, 367–380.

Panofsky, H. A., and G. W. Brier, 1968: *Some Applications of Statistics to Meteorology.* Penn State University Press, 224 pp.

Parrish, D. F., and J. C. Derber, 1992: The National Meteorological Center's spectral statistical-interpolation analysis scheme. *Mon. Wea. Rev.,* 120, 1747–1763.

Payne, S. W., 1982: A semi-prognostic study of a coupled cumulus and PBL parameterizations. *14th Tech. Conf. on Hurricanes and Tropical Meteorology,* San Diego, CA, Amer. Meteor. Soc.

Perkey, D. J., and C. W. Kreitzberg, 1976: A time-dependent lateral boundary scheme for limited-area primitive equation models. *Mon. Wea. Rev.,* 104, 744–755.

Peterson, K. R., 1966: Estimating low-level tetroon trajectories. *J. Appl. Meteor.,* 5, 553–564.

Phillips, N. A., 1979: The Nested Grid Model. NOAA Tech. Memo.

Pielke, R. A., 1974: A three-dimensional numerical model of the sea breeze over south Florida. *Mon. Wea. Rev.,* 102, 115–139.

——, 1984: *Mesoscale Meteorological Modeling.* Academic Press, 612 pp.

——, and T. Mahrer, 1975: Technique to represent the heated planetary boundary layer in mesoscale models with coarse vertical resolution. *J. Atmos. Sci.,* 32, 2288–2308.

——, and R. A. Avissar, 1990: Influence of landscape structure on local and regional climate. *Landscape Ecol.,* 4, 133–155.

——, G. Dalu, J. C. Snook, T. J. Lee, and T. G. F. Kittel, 1991: Nonlinear influence of mesoscale land-use on weather and climate. *J. Climate,* 4, 1053–1069.

——, W. R. Cotton, R. L. Walko, C. J. Tremback, M. E. Nicholls, M. D. Moran, D. A. Wesley, T. J. Lee, and J. H. Copeland,

1992: A comprehensive meteorological modeling system—RAMS. *Meteor. Atmos. Phys.*, **49**, 69–91.

Puri, K., and M. J. Miller, 1990: The use of satellite data in the specification of convective heating for diabatic initialization and moisture adjustment in NWP models. *Mon. Wea. Rev.*, **118**, 1081–1093.

Ramamurthy, M. K., and F. H. Carr, 1987: Four-dimensional data assimilation in the monsoon region. Part I: Experiments with wind data. *Mon. Wea. Rev.*, **115**, 1678–1706.

Randall, D. A., and Q. Shao, 1992: A second-order bulk boundary-layer model. *J. Atmos. Sci.*, **49**, 1903–1923.

Raymond, W. H., and R. B. Stull, 1990: Application of transilient turbulence theory to mesoscale numerical weather forecasting. *Mon. Wea. Rev.*, **115**, 2471–2499.

Reisinger, L. M., and S. F. Mueller, 1983: Comparisons of tetroons and computed trajectories. *J. Climate Appl. Meteor.*, **22**, 664–672.

Richtmyer, R. D., and K. W. Morton, 1967: *Difference Methods for Initial-Value Problems.* 2d ed., J. Wiley and Sons, 405 pp.

Ritter, B., and J.-F. Geleyn, 1992: A comprehensive radiation scheme for numerical weather prediction models with potential applications in climate simulations. *Mon. Wea. Rev.*, **120**, 303–325.

Rockel, B., E. Raschke, and B. Weyres, 1991: A parameterization of broad band radiative transfer properties of water, ice, and mixed clouds. *Beitr. Phys. Atmos.*, **64**, 1–12.

Rodgers, C. D., 1967: The use of emissivity in atmospheric radiation calculations. *Quart. J. Meteor. Soc.*, **93**, 43–54.

——, and C. D. Walshaw, 1966: The computation of infrared cooling rate in planetary atmospheres. *Quart. J. Meteor. Soc.*, **92**, 67–92.

Rogers, E., G. J. DiMego, J. P. Gerrity, R. A. Petersen, B. D. Schmidt, and D. M. Kann, 1990: Preliminary experiments using GALE observations at the National Meteorological Center. *Bull. Amer. Meteor. Soc.*, **71**, 319–333.

Rolph, G. D., and R. R. Draxler, 1990: Sensitivity of three-dimensional trajectories to the spatial and temporal densities of the wind field. *J. Appl. Meteor.*, **29**, 1043–1054.

Rosenthal, S. L., 1978: Numerical simulation of tropical cyclone development with latent heat release by the resolvable scales. I: Model description and preliminary results. *J. Atmos. Sci.*, **35**, 258–271.

——, 1979: The sensitivity of simulated hurricane development to cumulus parameterization details. *Mon. Wea. Rev.*, **107**, 193–197.

Rotunno, R., and K. A. Emanuel, 1987: An air–sea interaction theory for tropical cyclones. Part II: Evolutionary study using a non-hydrostatic axisymmetric numerical model. *J. Atmos. Sci.*, **44**, 542–561.

Sardie, J. M., and T. T. Warner, 1985: A numerical study of the development mechanisms of polar lows. *Tellus*, **37A**, 460–477.

Sasaki, K., H. Kurita, G. R. Carmichael, Y.-S. Chang, K. Murano, and H. Ueda, 1988: Behavior of sulfate nitrate, and other pollutants in the long-range transport of air pollution. *Atmos. Environ.*, **22**, 1301–1308.

Sasaki, Y., 1958: An objective analysis based on the variational method. *J. Meteor. Soc. Japan*, **36**, 1–88.

——, 1969: Proposed inclusion of time variation terms, observational and theoretical, in numerical variational objective analysis. *J. Meteor. Soc. Japan*, **47**, 115–124.

——, 1970a: Some basic formalisms in numerical variational analysis. *Mon. Wea. Rev.*, **98**, 875–883.

——, 1970b: Numerical variational analysis formulated under the constraints as determined by longwave equations and a low-pass filter. *Mon. Wea. Rev.*, **98**, 884–898.

Sasamori, T., 1968: The radiative cooling calculation for application to general circulation experiments. *J. Appl. Meteor.*, **7**, 721–729.

——, 1972: A linear harmonic analysis of atmospheric motion with radiative dissipation. *J. Meteor. Soc. Japan*, **50**, 505–517.

Sashegyi, K. D., and R. V. Madala, 1993: Application of vertical mode initialization to a limited area model in flux form. *Mon. Wea. Rev.*, **121**, 207–220.

——, D. E. Harms, R. V. Madala, and S. Raman, 1993: Application of the Bratseth scheme for the analysis of GALE data using a mesoscale model. *Mon. Wea. Rev.*, **121**, 2331–2350.

Savijärvi, H., 1990: Fast radiation parameterization schemes for mesoscale and short-range forecast models. *J. Appl. Meteor.*, **29**, 437–447.

Schlatter, T. W., 1988: Past and present trends in the objective analysis of meteorological data for nowcasting and numerical forecasting. Preprints, *Eighth Conf. on Numerical Weather Prediction,* Baltimore, MD, Amer. Meteor. Soc., J9–J25.

Schmidt, J. M., and W. R. Cotton, 1990: Interactions between upper and lower tropospheric gravity waves on squall line structure and maintenance. *J. Atmos. Sci.*, **47**, 1205–1222.

Schubert, W. H., 1974: Cumulus parameterization theory in terms of feedback and control. Atmospheric Science Paper No. 226, Colorado State University, 19 pp. [Available from Department of Atmospheric Sciences Building, Colorado State University, Fort Collins, CO 80523.]

Schumann, U., 1977: Realizability of Reynolds stress turbulence models. *Phys. Fluids.*, **20**, 721–725.

——, 1990: Large-eddy simulation of the up-slope boundary layer. *Quart. J. Roy. Meteor. Soc.*, **116**, 637–670.

Schwartz, B. E., and C. A. Doswell III, 1991: North American rawinsonde observations: Problems, concerns, and a call to action. *Bull. Amer. Meteor. Soc.*, **72**, 1885–1896.

Seaman, R. S., 1988: Some real data tests of the interpolation accuracy of Bratseth's successive correction method. *Tellus*, **40A**, 173–176.

Segal, M., and R. A. Pielke, 1981: Numerical model simulation of human biometeorological heat load conditions—Summer day case study for the Chesapeake Bay area. *J. Appl. Meteor.*, **20**, 735–749.

——, R. Avissar, M. C. McCumber, and R. A. Pielke, 1988: Evaluation of vegetative effects on the generation and modification of mesoscale circulation. *J. Atmos. Sci.*, **45**, 2268–2292.

——, W. E. Schreiber, G. Kallos, J. R. Garratt, A. Rodi, J. Weaver, and R. A. Pielke, 1989: The impact of crop areas in northeast Colorado on midsummer mesoscale thermal circulations. *Mon. Wea. Rev.*, **117**, 809–825.

Seltzer, M. A., R. E. Pasarelli, and K. A. Emanuel, 1985: The possible role of symmetric instability in the formation of precipitation rainbands. *J. Atmos. Sci.*, **42**, 2207–2219.

Shapiro, L. J., and H. E. Willoughby, 1982: The response of balanced hurricanes to local sources of heat and momentum. *J. Atmos. Sci.*, **39**, 378–394.

Shaw, D. B., P. Lonnberg, A. Hollingsworth, and P. Unden, 1987: Data assimilation: The 1984/85 revisions of the ECMWF mass and wind analysis. *Quart. J. Roy. Meteor. Soc.*, **113**, 533–566.

Shi, J. J., S. Chang, K. Sashegyi, and S. Raman, 1991: Enhancement of objective analysis of Hurricane Florence (1988) with dropsonde data. Preprints, *19th Conf. on Hurricanes and Tropical Meteorology,* Miami, FL, Amer. Meteor. Soc., 335–337.

Slingo, J. M., 1980: A cloud parameterization scheme derived from GATE data for use with a numerical model. *Quart. J. Roy. Meteor. Soc.*, **106**, 747–770.

——, 1987: The development and verification of a cloud prediction scheme for the ECMWF numerical model. *Quart. J. Roy. Meteor. Soc.*, **113**, 899–927.

Smagorinsky, J., K. Miyakoda, and R. F. Strickler, 1970: The relative importance of variables in initial conditions for dynamical weather prediction. *Tellus*, **22**, 141–157.

Somieski, F., P. Koepke, K. T. Kriebel, and R. Meerkötter, 1988: Improvements of simple radiation schemes for mesoscale models: A case study. *Beitr. Phys. Atmos.*, **61**, 204–218.

Soong, S.-T., and Y. Ogura, 1980: Response of tradewind cumuli to large-scale processes. *J. Atmos. Sci.*, **37**, 2035–2050.

Sorbjan, Z., 1986: On the similarity in the atmospheric boundary layer. *Bound.-Layer Meteor.*, **34**, 377–397.

——, 1987: An examination of local similarity theory in the stably stratified boundary layer. *Bound.-Layer Meteor.,* **38,** 63–71.

——, 1989: *Structure of the Atmospheric Boundary Layer.* Prentice Hall, 317 pp.

——, and M. Uliasz, 1982: Comparitive study of simple numerical models of the atmospheric boundary layer. *Acta Geophys. Pol.,* **30,** 167–176.

Stauffer, D. R., and N. L. Seaman, 1987: A real-data numerical study and four-dimensional data assimilation application for mesobeta-scale flow in complex terrain. *Proc. IUGG (IAMAP) Symp.,* Vancouver, British Columbia, Canada, IAMAP, 533–538.

——, ——, and F. S. Binkowski, 1991: Use of four-dimensional data assimilation in a limited-area mesoscale model. Part II: Effects of data assimilation within the planetary boundary layer. *Mon. Wea. Rev.,* **119,** 734–759.

Stein, U., and P. Alpert, 1993: Factor separation in numerical simulations. *J. Atmos. Sci.,* **51,** 2107–2115.

Stephens, G. L., 1978: Radiation profiles in extended water clouds. II: Parameterization schemes. *J. Atmos. Sci.,* **35,** 2123–2132.

——, 1979: Optical properties of eight water cloud types. Division of Atmospheric Physics Technical Paper No. 36, Commonwealth Scientific and Industrial Research Organization, 35 pp.

——, 1983: The influence of radiative transfer on the mass and heat budgets of ice crystals falling in the atmosphere. *J. Atmos. Sci.,* **40,** 1729–1739.

——, 1984: Review of the parameterization of radiation for numerical weather prediction and climate models. *Mon. Wea. Rev.,* **112,** 826–867.

Stephens, J. J., 1970: Variational initialization with the balance equation. *J. Appl. Meteor.,* **9,** 732–739.

Steyn, D. G., and I. G. McKendry, 1988: Quantitative and qualitative evaluation of a three-dimensional mesoscale numerical model simulation of a sea breeze in complex terrain. *Mon. Wea. Rev.,* **116,** 1914–1926.

Stocker, R. A., R. A. Pielke, A. J. Verdon, and J. T. Snow, 1990: Characteristics of plume releases as depicted by balloon launchings and model simulations. *J. Appl. Meteor.,* **29,** 53–62.

Straka, J. M., and J. R. Anderson, 1993: The numerical simulations of microburst producing thunderstorms: Some results from storms observed during the COHMEX experiment. *J. Atmos. Sci.,* **50,** 1329–1348.

——, and D. S. Zrnić, 1993: An algorithm to deduce hydrometeor types and contents from multiparameter radar data. Preprints, *26th Int. Conf. on Radar Meteorology,* Norman, OK, 513–515.

Stull, R. B., 1973: Inversion-rise model based on penetrative convection. *J. Atmos. Sci.,* **30,** 1092–1099.

——, 1984: Transilient turbulence theory. Part I: The concept of eddy mixing across finite distances. *J. Atmos. Sci.,* **41,** 3351–3367.

——, 1988: *An Introduction to Boundary Layer Meteorology.* Kluwer Academic, 666 pp.

——, 1993: Review of non-local mixing in turbulent atmospheres: Transilient turbulence theory. *Bound.-Layer Meteor.,* **62,** 21–96.

——, and A. G. M. Driedonks, 1987: Applications of the transilient turbulence parameterization to atmospheric boundary-layer simulations. *Bound.-Layer Meteor.,* **40,** 209–239.

Sugi, M., 1986: Dynamic normal mode initialization. *J. Meteor. Soc. Japan,* **64,** 623–636.

Sundqvist, H., 1978: A parameterization scheme for non-convective condensation including prediction of cloud and water content. *Quart. J. Roy. Meteor. Soc.,* **104,** 677–690.

Sundstrom, A., and T. Elvius, 1979: Computational problems related to limited-area modelling. Vol. 2, *Numerical Methods Used in Atmospheric Models,* GARP Publication Series No. 17, World Meteorological Organization, 379–416.

Talagrand, O., and P. Courtier, 1987: Variational assimilation of meteorological observations with the adjoint vorticity equation. Part 1: Theory. *Quart. J. Roy. Meteor. Soc.,* **113,** 1311–1328.

Tao, W.-K., J. Simpson, and S.-T. Soong, 1987: Statistical properties of a cloud ensemble: A numerical study. *J. Atmos. Sci.,* **44,** 3175–3187.

Temperton, C., 1988: Implicit normal mode initialization. *Mon. Wea. Rev.,* **116,** 1013–1031.

——, and D. L. Williamson, 1981: Normal mode initialization for a multilevel grid-point model. Part I: Linear aspects. *Mon. Wea. Rev.,* **109,** 729–743.

Tesche, T. W., 1988: Accuracy of ozone air quality models. *J. Environ. Eng.,* **114,** 739–752.

——, P. Georgopoulos, J. H. Seinfeld, P. M. Roth, F. W. Lurmann, and G. Cass, 1990: Improvements in procedures for evaluating photochemical models. CARB Report A823-103, California Air Resources Board, Sacramento, CA, 89 pp.

Therry, G., and P. Lacarrére, 1983: Improving the eddy kinetic energy model for planetary boundary layer description. *Bound.-Layer Meteor.,* **25,** 63–88.

Thiebaux, H. J., and M. A. Pedder, 1987: *Spatial Objective Analysis.* Academic Press, 299 pp.

Thuillier, R. H., 1992: Evaluation of a puff dispersion model in complex terrain. *J. Air Waste Management Assoc.,* **42,** 290–297.

Tiedtke, M., 1988: Parameterization of cumulus convection in large-scale models. *Physically Based Modelling and Simulation of Climate and Climate Change,* M. Schlesinger, Ed., Reidel, 375–431.

Tjernstrøm, M., 1987: A study of flow over complex terrain using a three-dimensional model: A preliminary model evaluation focusing on stratus and fog. *Ann. Geophys.,* **5B,** 469–486.

Tremback, C. J., G. J. Tripoli, R. Arritt, W. R. Cotton, and R. A. Pielke, 1986: The regional atmospheric modeling system. *Proc. Int. Conf. Development and Application of Computer Techniques to Environmental Studies,* Los Angeles, CA, Computational Mechanics Institute, 601–607.

Tripoli, G. J., 1986: A numerical investigation of an orogenic mesoscale convective system. Atmospheric Sciences Paper No. 401, 209 pp. [Available from Department of Atmospheric Sciences, Atmospheric Sciences Building, Colorado State University, Fort Collins, CO 80523.]

——, and W. R. Cotton, 1989a: A numerical study of an observed orogenic mesoscale convective system. Part 1: Simulated genesis and comparison with observations. *Mon. Wea. Rev.,* **117,** 273–304.

——, and ——, 1989b: A numerical study of an observed orogenic mesoscale convective system. Part 2: Analyses of governing dynamics. *Mon. Wea. Rev.,* **117,** 305–328.

——, C. Tremback, and W. R. Cotton, 1986: A comparison of a numerically simulated mesoscale system with explicit and parameterized convection. *23d Conf. on Cloud Physics and Radar Meteorology,* Snowmass, CO, Amer. Meteor. Soc., J131–J134.

Turpeinen, O. M., 1990: Diabatic initialization of the Canadian regional finite-element (RFE) model using satellite data. Part II: Sensitivity to humidity enhancement, latent-heating profile and rain-rate. *Mon. Wea. Rev.,* **118,** 1396–1407.

——, L. Garand, R. Benoit, and M. Roch, 1990: Diabatic initialization of the Canadian regional finite-element (RFE) model using satellite data. Part I: Methodology and application to a winter storm. *Mon. Wea. Rev.,* **118,** 1381–1395.

Uliasz, M., 1993: The atmospheric mesoscale dispersion modeling system. *J. Appl. Meteor.,* **32,** 139–149.

——, and R. A. Pielke, 1992: Effect of land surface representation on simulated mesoscale pollution dispersion. *Air Pollution Modeling and Its Application IX,* H. van Dop and G. Kallos, Eds., Plenum Press, 163–170.

Ulrickson, B. L., and C. F. Mass, 1990: Numerical investigation of mesoscale circulations over the Los Angeles basin. Part I: A verification study. *Mon. Wea. Rev.,* **118,** 2138–2161.

Venkatram, A., 1988: Model evaluation. *Lectures on Air Pollution Modeling,* Amer. Meteor. Soc., 308–321.

——, P. Karamchandani, and P. K. Misra, 1988: Testing a comprehensive acid deposition model. *Atmos. Environ.,* **22,** 737–747.

Vukicevic, T., and R. Errico, 1990: The influence of artificial and physical factors upon predictability estimates using a complex limited-area model. *Mon. Wea. Rev.,* **118,** 1460–1482.

Walko, R. L., W. R. Cotton, and R. A. Pielke, 1992: Large-eddy simulations of the effects of hilly terrain on the convective boundary layer. *Bound.-Layer Meteor.,* **58,** 133–150.

Walmsley, J. L., and J. Mailhot, 1981: On a method of evaluation of performance of a trajectory model for long-range transport of atmospheric pollutants. Vol. 2, *Air Pollution Modeling and Its Application,* C. DeWispelaere, Ed., Plenum Press, 277–284.

Wang, W., and T. T. Warner, 1988: Use of four-dimensional data assimilation by Newtonian relaxation and latent heat forcing to improve a mesoscale model precipitation forecast: A case study. *Mon. Wea. Rev.,* **116,** 2593–2613.

Warner, T. T., R. R. Fizz, and N. L. Seaman, 1983: A comparison of two types of atmospheric transport models—Use of observed winds versus dynamically predicted winds. *J. Climate Appl. Meteor.,* **22,** 394–406.

——, D. Keyser, and L. W. Uccellini, 1984: Some practical insights into the relationship between initial state uncertainty and mesoscale predictability. *Proc. Institute of Physics Conf.,* La Jolla, CA, American Institute of Physics, 271–286.

——, L. E. Key, and A. M. Lario, 1989: Sensitivity of mesoscale-model forecast skill to some initial data characteristics: Data density, data position, analysis procedure and measurement error. *Mon. Wea. Rev.,* **117,** 1281–1310.

Williamson, D. L., and P. J. Rasch, 1989: Two-dimensional semi-Lagrangian transport with shape-preserving interpolation. *Mon. Wea. Rev.,* **117,** 102–129.

Willmott, C. J., 1982: Some comments on the evaluation of model performance. *Bull. Amer. Meteor. Soc.,* **63,** 1309–1313.

——, S. G. Ackleston, R. E. Davis, J. J. Feddema, K. M. Klink, D. R. Legates, J. O'Donnell, and C. M. Rowe, 1985: Statistics for the evaluation and comparison of models. *J. Geophys. Res.,* **90**(C5), 8995–9005.

Wilson, N. R., and R. H. Shaw, 1977: A higher order closure model for canopy. *J. Appl. Meteor.,* **16,** 1197–1205.

Wippermann, F., 1971: A mixing-length hypothesis for the planetary boundary layer flow in the atmosphere. *Beitr. Phys. Atmos.,* **44,** 215–226.

Wolcott, S. W., and T. T. Warner, 1981: A moisture analysis procedure utilizing surface and satellite data. *Mon. Wea. Rev.,* **109,** 1989–1998.

Wyngaard, J. C., and O. R. Coté, 1974: The evolution of a convective planetary boundary layer. A higher-order closure study. *Bound.-Layer Meteor.,* **7,** 289–308.

Yamada, T., 1975: The critical Richardson number and the ratio of the eddy transport coefficients obtained from a turbulence closure model. *J. Atmos. Sci.,* **32,** 926–933.

——, 1982: A numerical model study of turbulent airflow in and above a forest canopy. *J. Meteor. Soc. Japan,* **60,** 439–454.

——, and S. Bunker, 1988: Development of a nested grid, second-moment turbulence closure model and application to the 1982 ASCOT Brush Creek data simulation. *J. Appl. Meteor.,* **27,** 562–578.

——, and ——, 1989: A numerical model study of nocturnal drainage flows with strong wind and temperature gradients. *J. Appl. Meteor.,* **28,** 545–554.

——, C.-Y. J. Kao, and S. Bunker, 1989: Air flow and air quality simulations over the western mountainous region with a four-dimensional data assimilation technique. *Atmos. Environ.,* **23,** 539–554.

——, S. Bunker, and S. Moss, 1992: Numerical simulations of atmospheric transport and diffusion over coastal complex terrain. *J. Appl. Meteor.,* **31,** 565–578.

Yamasaki, M., 1987: Parameterization of cumulus convection in a tropical cyclone model. *Short and Medium Range Numerical Weather Prediction,* T. Matsuno, Ed., Universal Academy Press, 665–678.

Yanai, M. S., S. Esbensen, and J. Chu, 1973: Determination of bulk properties of tropical cloud clusters from large-scale heat and moisture budgets. *J. Atmos. Sci.,* **30,** 611–627.

Zdunkowski, W. G., R. M. Welch, and G. Korb, 1980: An investigation of the structure of typical two-stream methods for the calculation of solar fluxes and heating rates in clouds. *Beitr. Phys. Atmos.,* **53,** 147–166.

Zhand, D.-L., 1989: The effect of parameterized ice microphysics on simulations of vortex circulation with a mesoscale hydrostatic model. *Tellus,* **41A,** 132–147.

——, and R. A. Anthes, 1982: A high resolution model of the planetary boundary layer—Sensitivity tests and comparisons with SESAME-79 data. *J. Appl. Meteor.,* **21,** 1594–1609.

——, E.-Y. Hsie, and M. W. Moncrieff, 1988: A comparison of explicit and implicit predictions of convective and stratiform precipitating weather systems with a meso-β-scale numerical model. *Quart. J. Roy. Meteor. Soc.,* **114,** 31–60.

Zhang, Y., M. Laube, and E. Raschke, 1992: Evolution of stratiform cirrus simulated in a lifting layer. *Beitr. Phys. Atmos.,* **65,** 23–33.

Zhao, Q., 1993: The incorporation and initialization of cloud water/ice in an operational forecast model. Ph.D. dissertation, University of Oklahoma.

Zilitinkevich, S. S., and D. L. Laikhtman, 1965: Turbulent regime in the surface layer of the atmosphere. *Izv. Acad. Sci. USSR, Atmos. Oceanic Phys.,* **1,** 150–156.

Zipser, E. J., 1977: Mesoscale and convective-scale downdrafts as distinct components of squall-line structure. *Mon. Wea. Rev.,* **105,** 1568–1589.

Zrnić, D. S., 1991: Complete polarimetric and Doppler measurements with a single receiver radar. *J. Atmos. Oceanic Technol.,* **8,** 159–165.